Owning the Genome

Owning the Genome

A Moral Analysis of DNA Patenting

David B. Resnik

State University of New York Press

Published by
State University of New York Press, Albany

© 2004 State University of New York

For information, address State University of New York Press,
90 State Street, Suite 700, Albany, NY 12207

Production by Michael Haggett
Marketing by Michael Campochiaro

Library of Congress Cataloging-in-Publication Data

Resnik, David B.
  Owning the genome : a moral analysis of DNA patenting / David B. Resnik
  p. cm.
  Includes bibliographical references and index.
  ISBN 0-7914-5931-4 (alk. paper) — ISBN 0-7914-5932-2 (pbk. : alk. paper)
    1. Patent laws and legislation—United States. 2. Biotechnology industries—Law and
legislation—United States. 3. Recombinant DNA—Patents—Moral and ethical aspects. 4.
Human genome—Patents—Moral and ethical aspects. I. Title.

KF3133.B56R47 2004
346.7304'86—dc22

                                                                        2004041648

10 9 8 7 6 5 4 3 2 1

For Susan, Peter, and Michael

Contents

Acknowledgments   ix

Legal Disclaimers   xi

Disclosure Statement   xiii

1 Introduction and Overview   1

2 DNA and Biotechnology   13

3 DNA as Intellectual Property   31

4 Arguments for DNA Patenting   63

5 Patenting Nature?   73

6 DNA Patents and Human Dignity   93

7 DNA Patents and Scientific Progress   131

8 DNA Patents and Medicine   155

9 DNA Patents and Agriculture   177

10 Conclusions and Policy Recommendations   195

Notes   203

References   211

Index   231

Acknowledgments

I would like to acknowledge individuals and organizations that have helped me gather information for this book or who have given me critical feedback. Those deserving recognition include: Ellen Clayton, Ronald Cole-Turner, Ken De Ville, John Doll, Rebecca Eisenberg, Lila Feisee, Myron Genel, Paul Gilman, Frank Grassler, Mark Hanson, Pamela Langer, David Magnus, Ron Mahurin, Jon Merz, Jeremy Rifkin, Todd Savitt, Richard Sharp, Jeremy Sugarman, John Wiley, Stephen Wilkinson, several anonymous reviewers, the National Human Genome Research Institute, and the U.S. Office of Patents and Trademarks. The ideas and opinions contained in this book do not necessarily reflect their views. Any mistakes of fact or reasoning are my own. I would also like to thank Kristy Aro, Georganne Perry, and Sarah Potter for help in preparing the manuscript.

Legal Disclaimers

The aim of this book is to provide the reader with some general information pertaining to legal issues. It does not aim to provide the reader with legal advice on any particular case. Those who require legal advice on intellectual property questions should consult a licensed attorney.

Since the aim of the book is to examine moral issues, legal scholars and patent attorneys may not find this book very useful for analyzing legal problems concerning biotechnology patents. Although the book discusses legal issues and principles, it aims only to provide an overall introduction to patent law for those who have an interest in the ethical and social questions related to DNA patents. My background is in philosophy: I earned my Ph.D. from the University of North Carolina in 1990 and have taught philosophy and ethics for over fifteen years. I undertook the study of law at Concord University while writing this book and completed my J.D. in 2003. Although I have some knowledge of the law, I consider myself to be a professor of philosophy and ethics, not an attorney or legal scholar.

Disclosure Statement

I have no significant financial interests in the companies discussed in this book nor have I received any funding from those companies in working on this book. I have no formal relationships with any of the companies discussed in this book.

1

Introduction and Overview

THE CONTROVERSY OVER PATENTING DNA

Since 1976, the United States' Patent and Trademark Office (PTO) has issued over 16,000 patents on isolated and purified deoxyribonucleic acid (DNA) sequences or on processes used to identify, isolate, copy, sequence, or analyze DNA sequences (PTO 2002). In 1999 alone, the office received over 3,000 patent applications pertaining to DNA sequences or DNA-related technologies (Enserink 2000). Although the practice of DNA patenting is scarcely more than a couple of decades old, it has created an enormous controversy. The storm began brewing in 1994, when the National Institutes of Health (NIH) applied for patents on thousands of gene fragments in an attempt to undercut private efforts to patent these DNA sequences. The PTO rejected these applications, however (Zurer 1994).

During that same year, over thirty organizations representing indigenous peoples announced formal declarations objecting to gene patenting, the ownership of life, and the commercial exploitation of indigenous peoples. These organizations were responding, in large part, to the NIH's patent applications on viral genes taken from the Hagahai people in Papua, New Guinea and natives of the Solomon Islands as well as the Human Genome Diversity Project, aka "the vampire project" (Taubes 1995; Crigger 1995). While the NIH's applications did not seek patents on human genes, these organizations nevertheless argued that the patents would harm and exploit indigenous peoples and violate their cultural values. Researchers claimed that these patents could deter scientific progress and that the NIH, a government agency, should not be involved in any attempt to exert proprietary control over DNA: the NIH

1

should encourage public dissemination of DNA and not private control. In defense of its patent applications, the agency claimed that it hoped to encourage private investment in the development of vaccines based on these viral genes and that it intended to grant nonexclusive licenses to companies. In 1995, the PTO awarded the NIH a patent on a viral gene taken from residents of Papua, New Guinea. Responding to objections from researchers as well as the public, the NIH withdrew its request to patent viral DNA sequences taken from residents of the Solomon Islands, although it retained the Papua, New Guinea patent at the request of clinicians working with that population, who felt that the population could benefit from research and royalties generated by the patent (Resnik 1999b).

In 1995, the PTO awarded the NIH and Genetic Therapy Incorporated patents on techniques for modifying cells ex vivo. Opponents of this patent argued that it was too broad and that it would stifle research (Beardsley 1994). The patent even drew Congress' ire, which considered but rejected a measure that would have prevented many types of gene patents (Kevles and Berkotwitz 2001). On May 18, of the same year, about 180 religious leaders, led by biotechnology critic Jeremy Rifkin, held a press conference in Washington, DC, objecting to biomedical patenting. In their "Joint Appeal against Human and Animal Patenting" (1995) these leaders denounced all attempts to patent nature. Some of the members of the Joint Appeal compared gene patenting to slavery, while others claimed that gene patents treat human beings as marketable commodities (Joint Appeal 1995; Andrews 1995; Peters 1997; Hanson 1997; Rifkin 1998). Religious leaders and organizations also took their mission to stop gene patenting beyond the Joint Appeal and published articles and editorials on the subject (Christian Century 1995; Land and Mitchell 1996). For Rifkin, the Joint Appeal vindicated ideas he had voiced for years. Since the 1970s, Rifkin has been the biotechnology industry's gadlfy. He has written books denouncing attempts to modify, engineer, patent, or own living things. His Foundation on Economic Trends, a Washington-based nonprofit organization, champions Rifkin's admonitions and concerns about the biotechnology revolution (Rifkin 1983, 1985, 1998, 2000).

From 1995–1999, the controversy continued as researchers, scholars, and government officials objected to private efforts to patents DNA technologies and DNA sequences. During this period, a handful of genomics companies, such as Celera Genomics, Human Genome Sciences, Genset, and Myriad Genetics were created with the explicit mission of marketing genetic information for use in diagnosis, therapy, and drug discovery. Their business plans called for DNA patenting, protein patenting, and the commercialization of genomics information services (Fisher 1999; Wade 2000a,b,c; Marshall 1999a,b,c; Wicklegren 1999). Many pharmaceutical and biotechnology companies, such as Incyte Pharmaceuticals, Glaxo Welcome, Millenium Pharmaceuticals, Genentech, Perkin Elmer, and Monsanto, also took an interest in gene patenting and bought gene patents or reached licensing agreements with companies conducting genetic research.

Private efforts to profit from genomics research were part of a massive increase in private funding of biomedical research and development (R & D) that has taken place in the last two decades. Private funding of biomedical R & D rose from $2 billion per year in 1980 to over $50 billion per year in 2000 (Beardsley 1994; Resnik 1999a). Although government funding still plays an important role in biomedical R & D, private funds now account for more than 60 percent of all biomedical R & D, including a large portion of genomics R & D (Resnik 1999a). As money continued to pour into privately funded genomics research, many people in the research community worried that private efforts to patent or control DNA would hamper scientific innovation and discovery (Marshall 1997; Heller and Eisenberg 1998; Reynolds 2000; Gosselin and Jacobs 2000; Guenin 1996; Caplan and Merz 1996; Merz et al. 1997).

Many scientists, clinical researchers, and organizations continue to oppose various types of DNA patenting. For example, the Council for Responsible Genetics drafted a Genetic Bill of Rights that opposes the patenting of human genes (Council for Responsible Genetics 2000). The United Nations Educational, Scientific and Cultural Organization (UNESCO) declared its opposition to human gene patenting several years ago (UNESCO 1997). Many indigenous groups have signed formal declarations against animal or human DNA patenting (Resnik 1999b). The Foundation on Economic Trends remains firmly opposed to all forms of gene patenting (Rifkin 2000).

Early debates about DNA patenting focused on fundamental questions about whether any patenting of DNA should be legal. While many people still oppose all forms of DNA patenting, these arguments have so far not swayed legislators, judges, or patent offices. Accordingly, debates about gene patenting have shifted away from general concerns about patenting toward more specific issues related to patenting (Barton 2000; Resnik 2001a,b; Caulfield and Gold 2000a,b; Heller and Eisenberg 1998). Many people have raised objections to specific patenting policies, such as allowing patents on sequences tags (ESTs) or single nucleotide polymorphisms (SNPs); patenting the use of genes to diagnose diseases; and patenting genes related to research on the human immunodeficiency virus (HIV) (American Society of Human Genetics 1991; Human Genome Organization 1995; Marshall 1997; Council on Ethical and Judicial Affairs 1997; Reichhardt 1998;Reynolds 2000; Marshall 2000a). Others have objected to the effects of patents on agricultural biotechnology and global trade (Poland 2000; Barton and Berger 2001; Shiva 1996). Most of these specific concerns address potential, undesirable social consequences of patenting and restrictions on access to genetic information. These critics argue that some types of DNA patents may hinder the progress of science, medicine, or agriculture.

Many government researchers, such as Francis Collins, Director of the National Human Genome Research Institute (NHGRI), which funds the Human Genome Project (HGP), have expressed dire concerns about private control of genetic information (Marshall 2000b,c,d,e). In response to concerns raised by researchers about speculative and broad patents on DNA sequences, PTO decided to raise the bar on DNA patents by clarifying the conditions that must be satisfied before a patent may be awarded. The PTO issued new utility guidelines in December 1999, which made it clear that inventors must state definite, specific, and plausible uses for the sequences of DNA that they plan to patent (Patent and Trademark Office 1999; Enserink 2000; Resnik 2001a).

During the 1990s, a rivalry developed between public and private efforts to conduct genomics R & D. The human genome, the holy grail of biomedicine, occupied ground zero in this conflict. The two principal players were Collins and Craig Venter, the (then) maverick CEO of Celera Genomics, which launched a private effort to sequence and map the human genome (Kevles and Berkowitz 2001). The public effort consisted of a consortium of universities and research centers, led by Collins and the NHGRI. Using a shotgun approach to gene mapping and sequencing as well as supercomputers and automated sequencing machines, Venter promised to sequence the human genome ahead of the HGP's schedule (Marshall 1999b; 2000b,e). Traditional clone-by-clone methods, used by researchers in the HGP, sequence DNA one sequence (or clone) at a time. The shotgun method breaks DNA into its various parts, sequences them all at once, and then uses supercomputers to reassemble the parts. Many scientists doubted whether his shotgun approach would work and feared that it would produce a genome with missing pieces or pieces that are out of place. Venter was vindicated in February 2000, when Celera produced a high-quality data set for the entire genome of the fruit fly, *Drosophila melanogaster* (Pennisi 2000a,b; Adams et al. 2000).

Access to genetic information was one of the key issues in this public-private conflict. The NHGRI wanted all researchers to have free access to genetic data as soon as it is checked for accuracy and quality; Celera planned to eventually allow all researchers free access to data through its website but would also charge corporations or research institutions a fee for an early look at genetic data. Although Celera did not plan to patent large portions of the genome, it planned to patent some specific genes or DNA sequences with practical applications. Celera planned to make most of its money through selling information services related to its genomics databases. (This business strategy proved to not be very profitable.) The NHGRI, on the other hand, hoped to undercut private patenting efforts by placing as much genetic information as possible in the public domain (Marshall 2000b,e).

In June 2000, the two sides agreed to cooperate and announced that the entire human genome had been sequenced and would soon be mapped and analyzed (Wade 2000c). They used data generated by both the clone-by-clone and shotgun methods to complete this task. In February 2001, the public consortium and Celera published versions of the human genome in the journals *Nature* and *Science*, respectively (Marshall 2001; Venter et al. 2001). Under the terms and conditions negotiated between *Science* and Celera, nonprofit researchers are allowed to download pieces of Celera's DNA sequence from its website, provided that they agree not to commercialize or distribute the data. Researchers who plan to use the data for commercial purposes must negotiate an agreement with Celera (Marshall 2001).

As an epilogue to this story, it is worth noting that Venter resigned as CEO of Celera in January 2002, to form another company, because he disagreed with the company's new business strategy. Celera is now pursuing drug development instead of selling access to data, as its main business objective. On August 16, 2002, Venter announced his plans to form a genome sequencing company, which will sequence human, animal, and plant DNA for a fee. He hopes that in ten years advances in sequencing technologies will allow his company to be able to sequence a person's entire genome for several thousand dollars, which would represent a dramatic cost-reduction. Currently, the cost of sequencing a human genome would run at least into the millions of dollars. Unlike his earlier company, Venter's new company will be nonprofit. Venter also now plans to release all data to the public at no cost. He has also established two foundations, the Center for the Advancement of Genomics, which explores ethical and policy issues related to genomics, and the Institute for Biological Energy Alternatives, which develops microorganisms that produce energy alternatives to fossil fuels (Pollack 2002).

The United States was not the only country debating DNA patenting. In Europe, researchers and the public also expressed concerns about patenting genes and life forms (Kevles and Berkowitz 2001). In France, legislators adopted a measure declaring that human genes (in their natural state) are not patentable, which brought the country in conflict with other members of the European Union (EU) (Balter 2000). Denmark, on the other hand, ruled that there is no compelling argument against gene patenting (Knoppers 1999). Switzerland considered but ultimately rejected an initiative that would have forbidden the patenting of transgenic plants, animals, or their components parts (Schatz 1998).

The European Commission, a division of the EU which makes policy for the European Patent Convention (EPC), ruled that the EPC can refuse to award patents on inventions that infringe on human rights or violate human dignity (European Commission 1998). The EPC's patent laws, unlike U.S. patent laws, declare that patents should not be granted for inventions that are contrary to the public morality (Brody 1999). Although gene patents remain legal in countries that accept the EPC, it is not clear whether gene patents are contrary to the public morality or violate human dignity (Crespi 2000).

OVERVIEW OF THE ISSUES

Although DNA patenting is a highly technical topic far removed from ordinary, human concerns, it is not entirely surprising that it has generated so much controversy, since a great deal is at stake in this issue. As mentioned earlier, private companies have invested billions of dollars in genetic research with the expectation that they will be able to obtain the intellectual property protection afforded by patents. Genomic R & D has had a significant impact on the world's economy and has played a key role in advances in pharmaceuticals, medicine, biotechnology, and agriculture (Rifkin 1998; Enriquez 1998; Biotechnology Industry Organization 2001a). It was clear that industrial biotechnology had come of age when the announcement by then President William Clinton of the United States and Prime Minister Tony Blair of the United Kingdom (UK) of an agreement to make data from the human genome available to all researchers sent the NASDAQ composite tumbling over 200 points (Berenson and Wade 2000). The NASDAQ contains many biotechnology companies, including Celera, and investors speculated that this announcement would undercut private efforts to profit from genomics. During the slide, Celera's stock dropped 5.2 percent and Incyte's dropped 12 percent. As an aside, Wall Street probably overreacted to this announcement because it only restated previous commitments by these governments to make genetic data publicly available, and biotechnology companies have taken these commitments into account in their business plans (Langreth and Davis 2000). However, the episode shows that even if investors misunderstood the significance of the announcement, proprietary interests in DNA and other biological materials have important implications for business and investing. Moreover, disputes about intellectual property rights in biotechnology can exacerbate the high volatility of biotech stocks.

The issues involve much more than money, however. Genetics and genomics are foundational disciplines in many different areas of biology, biotechnology, and agriculture, and have an important bearing on psychology, sociology, and anthropology (Kitcher 1997). The free and open exchange of information is vital for discovery and innovation in basic and applied research (Resnik 1998b,c). Practices that can inhibit access to genetic data and genetic technologies, such as patenting, can therefore be an impediment to the progress of science. Likewise, genetics and genomics now play a vital role in diagnosing, treating, and preventing human diseases and in agricultural biotechnology (Collins and McKusick 2001; Barton and Berger 2001). Now that researchers have completed their study of the human genome, it will be possible to understand the genetic basis of health and disease. Still, this genetic information is of little use if it is not available to clinicians and medical researchers. If gene patenting were to restrict access to genetic information and genetic technologies, then this would threaten the progress of medicine and the promotion of health. The same points apply to restrictions on access to genetic information that could affect agriculture. Thus, basic researchers, applied scientists, and clinicians are key stakeholders in the gene patenting controversy.

Genetics and genomics also have important implications for society and culture. Since the discovery of the structure of DNA in 1953, genes have acquired a great deal of cultural and social significance and symbolic value (Nelkin and Lindee 1995). Since many people readily accept genetic explanations of human personality, behavior, and physiology, and genetic tests can be used in medicine, insurance, employment, or the criminal justice system, our treatment of genes has ramifications for human rights, privacy, and dignity. Although few people would equate a person with a set of genes, many people believe the genes have some fundamental connection to the person or self. Since genetics and genomics play key roles in biology, medicine, and agriculture, our attitudes toward genes and DNA have implications for our views of humankind's relation to nature. Since researchers may use DNA sequences from all over the world, genetics and genomics also have important implications for cultural exploitation. Furthermore, since genetically engineered crops or animals could affect the health and safety of other species, including humans, genetics and genomics have important implications for public health and the environment. Thus, virtually all people who are concerned with the cultural, social, and environmental consequences of bioscience and biotechnology, such as religious leaders, politicians, environmentalists, consumers, humanists, indigenous people, and ethicists also have a stake in the DNA patenting dispute.

DNA patenting raises a number of different legal, ethical, philosophical and political issues. These questions range from very broad concerns about public policy, such as "What is the justification of the patent system?" and "What is the difference between a product of nature and a product of human ingenuity?" to narrower questions, such as, "Are isolated and purified DNA sequences products of human ingenuity?" and "What is the correct way to interpret the scope of a DNA patent?" Although most of these questions are framed in terms of specific laws or policies and have a legal context, in order to answer them, one must often examine the moral and philosophical arguments used to justify policies and frame legal questions and issues. For example, U.S. statutes and court decisions provide a basis for the legal distinction between a product of nature and a product of human ingenuity, but to interpret or evaluate this distinction one must have some grasp of its moral and philosophical implications. Although the courts operate according to legal rules and procedures, they must frequently interpret concepts and terms from ordinary language and address public policy arguments. The line between law and ethics that seems so clear in the abstract becomes quite murky when one examines a real world legal issue with moral, social, and economic implications. Although this book does not seek to render a legal analysis of DNA patenting or offer legal advice, it will be necessary to engage the legal issues that raise moral and ethical concerns. One goal of this book is to provide a moral analysis of DNA patenting that can be used to interpret patenting laws or to suggest changes in the laws.

MY APPROACH: A MORAL ANALYSIS

What makes an analysis of a practical problem a *moral* analysis as opposed to some other type of analysis, such as a political, legal, or economic analysis? A moral analysis, according to many, attempts to understand whether a particular action or policy is right or wrong, all things considered (Fox and DeMarco 1990; Baier 1958). It attempts to determine whether there are good reasons for performing an action or instituting a policy. Thus, a moral analysis may consider and critique economic, legal, political, social, religious, and scientific perspectives related to the particular question at hand. A moral analysis of a practical problem takes all relevant factual and normative considerations into account, considers the relevant interests at stake, examines the issue from different perspectives, and attempts to reach a fair and impartial decision (Rachels 1993). It attempts to give a well-reasoned analysis of a practical problem, and involves the kind of self-reflective and critical discussion associated with the Socratic method used in philosophical debate and legal argument. Thus, to understand whether DNA patenting is morally justifiable, one must address the economic, scientific, social, political, legal, and religious aspects of the issue.

There are two basic viewpoints one may consider when conducting a moral analysis (Frankena 1973; Pojman 1995). According to consequentialist approaches to morality, an action or policy is justifiable insofar as it is likely to yield the greatest balance of good/bad consequences (outcomes or results) for all relevant parties. Utilitarianism is the most influential consequentialist theory in ethics. According to this view, one should act so as to maximize utility and minimize disutility. There is considerable debate among utilitarians about how one should define or measure utility. Early utilitarians, such as John Stuart Mill ([1861]1979), equated utility with happiness; modern utilitarians define utility in terms of welfare or the satisfaction of preferences (Scheffler 1988).

Consequentialist approaches also play a prominent role in economic theory, environmental management, medicine, and public health. Economists frequently analyze policies, technologies, and institutional arrangements in terms of their economic costs and benefits (Blaug 1980; Samuelson 1980). Scholars and policy analysts concerned with environmental considerations or public health and safety often use risk-benefit theory to understand the consequences of actions or policies. Risk-benefit theories address the probability and magnitude of harms as well as the probability and magnitude of benefits (Shrader-Frechette 1991). In medicine, physicians use evidence from physical examinations, medical records, diagnostic tests, and scientific articles to assess probable risks and benefits to the patient when making medical recommendations and decisions (Sackett et al. 1997).

Nonconsequentialist (or deontological) approaches, on the other hand, hold that the morality of an action or policy does not depend on its consequences: an action or policy is, by its very nature, moral or immoral, just or unjust (Frankena 1973; Pojman 1995). Kantianism is by far the most influential deontological theory in ethics. According to the eighteenth century German philosopher, Immanuel Kant, it is wrong to treat people as mere instruments to other ends, regardless of the consequences (Kant 1981 [1985]). Another popular deontological theory known as libertarianism, holds that all people are endowed with some basic natural rights to life, liberty, and property (Nozick 1974). The sole function of the state is to protect these rights, and restrictions on natural rights are justified only to prevent people from violating each other's rights.

Many theorists, such as Rawls (1971) and Ross (1930), defend a hybrid approach and hold that one needs to balance consequentialist and deontological concerns in determining how one should act. According this balancing approach, one must weigh and consider many different moral duties and principles in order to make a moral choice. It is not my aim in this book to pass judgment on consequentialist or deontological approaches to ethics and moral reasoning. I agree with commentators who recognize that both approaches should play some role in social and political philosophy and moral argument (Rawls 1971; Feinberg 1973; Pojman 1995; Gutman and Thompson 1996). A moral analysis of a problem should consider deontological concerns pertaining to moral rights, duties, and justice and consequentialist concerns relating to probable benefits, harms, and utility. One should also consider both of these perspectives in understanding the morality of DNA patents (Resnik 1997).

In this book, I shall examine the main arguments for and against DNA patenting from both consequentialist and deontological perspectives. (Table 1.1 provides an outline of these arguments.) I will examine and critique deontological arguments for and against DNA patenting and show that they generally fail to show that DNA patenting is inherently moral or inherently immoral. Only one type of DNA patenting is inherently immoral, the patenting of a whole human genome. The morality of all other forms of DNA patenting, from the patenting of gene markers, to whole genes, to artificial chromosomes, depends on the consequences of these practices for science, medicine, agriculture, society, business, industry, and the economy.

I shall attempt to show that DNA patenting offers society many important benefits, even though it also creates some potential threats. Although it

TABLE 1.1
Arguments for and against DNA Patenting

	Consequentialist Arguments	Deontological Arguments
For	Patenting promotes science and technology	People have a right to patent DNA
	Patenting benefits business and industry	
	Patenting benefits medicine	
	Patenting benefits agriculture	
Against	Patenting hinders science and technology	Patenting violates human dignity
	Patenting harms medicine	DNA is God's invention
	Patenting harms agriculture	DNA is our common heritage
	Patenting harms society and culture	Patenting DNA commodifies nature

is easy to imagine and predict the various ways that DNA patents might harm society, it is not easy to assign objective probabilities to these threats. In order to suggest strategies for decision-making when we lack a great deal of knowledge about potential outcomes, I shall articulate and defend a popular principle known as the Precautionary Principle (PP). I shall apply this principle to the DNA patenting debate and argue that the most reasonable response to the various threats posed by DNA patenting is to enact various regulations and policies to regulate DNA patenting. These rules would aim to prevent potential threats from happening or minimize their impact. We should take advantage of the opportunities presented by industrial biotechnology, but we should also take precautionary measures to mitigate harmful results. Throughout this book, I will discuss policies that I think would be reasonable responses to the potential harms of DNA patenting, and I will summarize my policy conclusions in the final chapter. Since the biotechnology industry is still in its infancy and DNA patenting is quite new, we need to study and monitor the effects of various policies and update them in response to changes in science, technology, and the industry (Resnik 2001a; Schonmann 1998). I will argue that there is no need, at this time, to make any substantive changes in patenting laws, since society already has the legal and regulatory tools needed to deals with the issues raised by DNA patenting.

Since this book will take a stand on a fairly controversial topic, I realize that many readers may disagree with my analysis of the problems and my proposed solutions. I do not aim to convince every reader of the truth of my views, but I do hope that even those who disagree with me will learn something from this book, that they will understand how we disagree, and that they will see where further discussion and debate may advance social policy.

2

DNA and Biotechnology

Before tackling the topic of DNA patenting, it will be useful to provide the general reader with some scientific background relating to the structure and function of DNA and the practical applications of biotechnology. Since I am only presenting introductory material for a lay audience, readers who are familiar with DNA and biotechnology may skip ahead to the next chapter.

FROM GENES TO DNA

The science of genetics began when the Austrian monk Gregor Mendel (1822–1884) conducted breeding experiments on inheritance in peas. Mendel bred plants with different traits, such as short versus tall plants and round seeds versus wrinkled seeds, and discovered patterns of inheritance (or Mendelian ratios) across generations. For example, he found that if he bred a plant with rounds seeds with one with wrinkled seeds, that these produced plants with round seeds. If he bred these plants, the next generation had a ratio of three round-seed plants to one wrinkled-seed plant. To explain these and other observations, Mendel posited the existence of unobserved factors (now known as genes or genotypes) that are transmitted from one generation to the next. Combinations of these factors produce different characters (now known as traits or phenotypes). Some of these factors are dominant while others are recessive. When a dominant factor and a recessive factor combine, the dominant factor (such as round seeds) will determine the trait found in the plant. Factors also randomly assort during the production gametes (i.e., pollen and eggs in peas or sperm and eggs in mammals). Unfortunately, most

13

scientists ignored Mendel's work, which was not appreciated until the early 1900s, when DeVries, Correns, Hunt, Morgan, Hardy, Weinberg, Fisher and other scientists developed a discipline based on his work known as Mendelian genetics (Mayr 1982; Ayala 1982).

During the 1900s, this new science became integrated into the established sciences of cytology, embryology, pathology, immunology, and physiology. Scientists sought to understand the mechanisms of inheritance in cells and the relationship between genetics and development. (Cells are the building blocks of all higher life forms and carry out the functions and activities of living things. Even simple organisms, such as the hydra, are composed of thousands of cells. Human beings are composed of billions of cells.) Geneticists developed and applied Mendel's laws to different species, and also found some exceptions to those laws, such as codominance (which violates the requirement that genes are either dominant or recessive) and gene linkage (which violates random assortment) (Ayala 1982). From the 1900s–1920s, scientists discovered that structures in the cell nucleus known as chromosomes carry genetic information. They learned how chromosomes line up and divide prior to cell division, and they distinguished between diploid organisms, which have two sets of chromosomes, and haploid organisms, which have only one set. By the 1930s, biologists had described and explained many cell structures and functions and the mechanisms of reproduction and development, but they still did not know how chromosomes divide and transmit genetic information. They also did not understand how the genetic information contained in chromosomes influences the activities of cells, tissues, and entire organisms (Mayr 1982).

During the 1940s and early 1950s, biologists began to learn more about the nucleic acids, deoxyribonucleic acid (DNA) and ribonucleic acid (RNA). The field of molecular biology emerged from this attempt to understand the biochemical basis of genetics and inheritance. Although biologists knew that chromosomes contain DNA and proteins, they still did not know how chromosomes carry genetic information and how they are able to divide. They also did not know whether genetic information is contained in DNA, RNA, or proteins. In 1953, James Watson (1928–) and Francis Crick (1916–), developed a model of the structure of DNA that provided the key to understanding how cells are able to encode and transmit genetic information. Using data from X ray crystallography and models of chemical bonding, Watson and Crick demonstrated that DNA is a long macromolecule consisting of two complementary strands composed of four deoxyribonucleotide pairs (or nucleotides). Since the molecule takes a helical shape, DNA resembles a winding, double helix (see figure 2.1). The four nucleotides are deoxyguanosine monophosphate (G), deoxycytidine monophosphate (C), deoxyadenosine monophosphate (A), and thymidine monophosphate (T). The nucleotide bases form different pairs: A pairs with T and G pairs with C. Thus, the

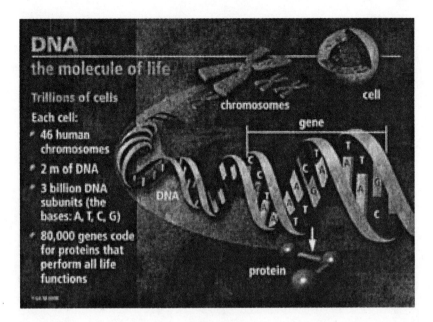

DNA
the molecule of life

Trillions of cells

Each cell:
- **46 human chromosomes**
- **2 m of DNA**
- **3 billion DNA subunits (the bases: A, T, C, G)**
- **80,000 genes code for proteins that perform all life functions**

cell

chromosomes

gene

DNA

protein

FIGURE 2.1. DNA—The Molecule of Life
Used by permission of the U.S. Department of Energy Human Genome Program,
http://www.ornl.gov/hgmis.

nucleotide sequence, AATTCG, would pair with the complementary
sequence, TTAAGC. As a result of base-pairing, the two strands of DNA in
the double-helix are mirror images of each other (Mayr 1982; Robertson
1994; Human Genome Project 2001).

Almost all organisms on this planet, with the notable exception of RNA-
viruses, encode their genetic information in DNA. An organism's genome is
the sum total of its DNA. The vast majority of DNA is housed in chromosomes, although mitochondria also contain a small amount of DNA
(MtDNA) (Kitcher 1997). Chromosomes are composed of proteins interwo-
ven with strands of DNA. The long strands of DNA usually exist as spaghetti-
like structures in the cell nucleus but condense prior to cell division and
become tightly wound in chromosomes. During cell division, DNA replicates
itself when an enzyme known as DNA-polymerase causes the two complementary strands to separate and bond with other nucleotides present in the
cytoplasm to form the missing, complementary strands (see figure 2.2). After
the DNA has replicated, the cell begins to split in half, and each daughter cell
receives a copy of the complete genome (Human Genome Project 2001).

While most genetic information is transmitted from one generation to the
next without any changes, modifications in the DNA, or mutations, frequently

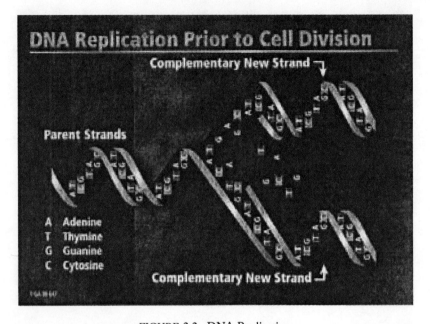

FIGURE 2.2. DNA Replication
Used by permission of the U.S. Department of Energy Human Genome Program,
http://www.ornl.gov/hgmis.

occur. Some mutations are caused by exposure to chemicals or radiation; some
are caused by diseases; and some result from errors in DNA replication. Mutations are natural and not always harmful. Indeed, evolution by natural selection cannot occur without them. It is also possible to artificially produce
mutations through genetic engineering, as we shall soon see. There are four
basic types of mutations: deletions, which involve the removal of one or more
base pairs; insertions, which involve the addition of one or more base-pairs;
substitution, which is the replacement of one or more base-pairs by other
base-pairs; and transposition, which is a change in the order of one or more
base-pairs (Ayala 1982). Although most mutations are benign, some result in
genetic diseases, such as cystic fibrosis (CF) or sickle cell anemia (SCA),
which we will soon discuss. Even a small mutation, such as the insertion of
one base pair, can result in a frame shift in the DNA, which can have dramatic
effects, because the DNA frame is the sequence of amino acids that encodes
a protein (Robertson 1994). For example, if a T is inserted in the sequence,
TAAGCG . . . A, then this would shift the frame to TTAAGC . . . A, which
could alter the protein product. While all cells have DNA repair mechanisms
whose function is to locate and correct mutations, mutations are a fact of life
(Wood et al. 2001).

The human genome consists of about three billion base-pairs. There are roughly 30,000–40,000 different genes in the human genome contained in 23 pairs of chromosomes, which consist of 22 pairs of nonsex chromosomes (or autosomes) and one pair of sex chromosomes, the X chromosome and the Y chromosome (Claverie 2001; Human Genome Project 2001). An average gene consists of about 3,000 base pairs but some genes have over 2 million base-pairs. An average chromosome contains 150,000 base-pairs (Kitcher 1997; Human Genome Project 2001). Human beings share approximately 99.9 percent of their DNA, so there is genetic variation in only 0.1 percent of human DNA (Subramanian et al. 2001). Only about 2 percent of the DNA in the human genome codes for proteins (Human Genome Project 2001). The regions of the genome that do not code for proteins, the noncoding regions, are also sometimes called "junk" DNA. Noncoding regions contain repetitive sequences of DNA, and the number of repeats has increased during evolutionary history. Even if they do not have biological functions, junk DNA sequences can still be useful in DNA typing (aka DNA fingerprinting) and in studying genetic diversity (Kitcher 1997). Scientists can study genetic diversity by identifying and analyzing shorter sequences of DNA (or gene fragments) known as single nucleotide polymorphisms (SNPs) (see figure 2.3).

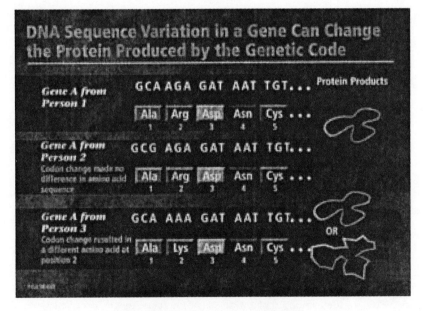

FIGURE 2.3. DNA Sequence Variation
Used by permission of the U.S. Department of Energy Human Genome Program, http://www.ornl.gov/hgmis.

Many of the SNPs can be found in the noncoding regions of the genome (Venter et al. 2001). One can also use haplotypes (unique combinations of genes) to study genetic diversity (Gabriel et al. 2002).

One can use SNPs to compare two different DNA samples to determine whether the samples are from the same individual or are from individuals that are closely genetically related (Paabo 2001; Venter et al. 2001). Indeed, the analysis of genetic variation across species indicates that human beings share about 98 percent of their DNA with chimpanzees and as much as 10 percent of their DNA with fruit flies. Although genetic changes provide the raw material for evolution by natural selection, much of the DNA found in eukaryotes (i.e., species that have a cell nucleus bound by a membrane) has been conserved during the two billion years that life has existed on the earth (Paabo 2001). Evolution by natural selection occurs when random, heritable variations lead to differential reproduction and survival among organisms in the same population in a given environment. Organisms that are best adapted to that environment are most likely to survive and reproduce in that environment (Ayala 1982; Mayr 1982). Thus, the fittest organisms tend to survive. Genetics is important in evolution because most traits are inherited through the genes, although some traits are inherited via other mechanisms, such as culture.

GENE EXPRESSION

RNA is macromolecule that plays a key role in gene expression, i.e. the process of manufacturing proteins from genetic information contained in DNA (see figure 2.4). RNA is single-stranded polymer consisting of the nucleotides adenine (A), uracil (U), guanine (G), and cytosine (C). These nucleotides form the same kinds of pairs as occur in DNA, except uracil takes the place of thymine. Thus, the DNA sequence AATTGCG pairs with the RNA sequence UUAACGC. During a process known as transcription, RNA polymerase triggers a reaction where an RNA molecule forms one nucleotide at a time as strands of DNA separate and then rejoin. Specific sequences of DNA, known as promoters and terminators, tell the RNA polymerase to start or stop this process. In this way, the information contained in DNA can be transcribed into messenger RNA (mRNA). The mRNA reads the sequences in the DNA frame. During translation, two other types of RNA, transfer RNA (tRNA) and ribosomal RNA (rRNA) decode the information contained in mRNA to form proteins (Robertson 1994; Kitcher 1997).

Proteins are composed of amino acids. There are twenty different types of amino acids normally found in proteins. Each amino acid corresponds to a nucleotide triplet (or codon). For example, the codon UUU corresponds to phenylalanine, and the codon AAA corresponds to lysine. During translation,

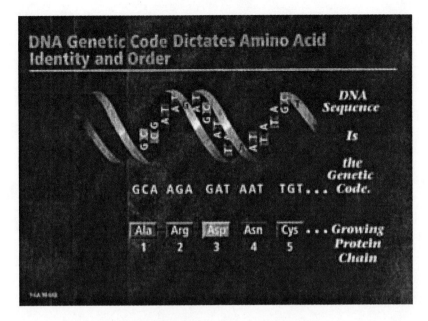

DNA Genetic Code Dictates Amino Acid Identity and Order

GCA AGA GAT AAT TGT... *the Genetic Code.*

| Ala | Arg | Asp | Asn | Cys | ... *Growing* |
| 1 | 2 | 3 | 4 | 5 | *Protein Chain* |

DNA Sequence Is the Genetic Code.

FIGURE 2.4. Genes and Proteins
Used by permission of the U.S. Department of Energy Human Genome Program,
http://www.ornl.gov/hgmis.

the rRNA housed in a ribosome reads the mRNA codons to form a protein one amino acid at a time. tRNA assists this process by forming the appropriate polypeptide bonds in the protein. The whole process stops when rRNA reaches a codon that tells it to stop. The protein can then be released into the cell and modified by other processes (Suzuki and Knudtson 1989; Kitcher 1997). If a single gene contains 3,000 base-pairs, then the corresponding protein will be composed of 1,000 amino acids. The total of all the different proteins in an organism is called the proteome (Human Genome Project 2001). There are as many as 2 million different proteins in the human body (Service 2001). There are more proteins than genes because cells alter the secondary, tertiary, and quaternary structures of amino acids once they are produced. Thus, one gene may encode information for making or modifying more than one protein.

By providing instructions for the sequence of amino acids, DNA contains information for the primary structure of a protein. A gene is therefore usually defined as the information contained in DNA required to make a single protein, which includes the DNA sequence that codes for the protein's primary structure as well as other sequences, such as promoters and terminators (Wong 1997). In diploid organisms, two copies of the same gene can play a role in production of a normally functioning protein. When a person has two mutated alleles (or versions of a gene), they may be unable to manufacture a normally functioning protein. For example, a person who has only one normal SCA allele carries the disease but does not have the disease. The individual can still produce a protein known as hemoglobin that functions normally, although the protein is somewhat affected by the presence of the mutated gene. However, if an individual lacks both normal alleles (and has two mutated alleles), then that individual will be unable to produce the protein properly and will have a genetic disease as a result (see figure 2.5). For instance, individuals who have SCA do not have the gene that plays a key role in the production the hemoglobin protein. As a result, they have abnormal hemoglobin and their red blood cells have a decreased capacity to carry oxygen. In some genetic diseases, such as Huntington's chorea, it only takes one mutated allele to cause abnormal gene expression (Resnik, Steinkraus, and Langer 1999).

Since the sequence of amino acids (or polypeptide chain) can bond with itself in many different ways, proteins have complex three-dimensional structures. For example, a polypeptide may form a coil (secondary structure), which

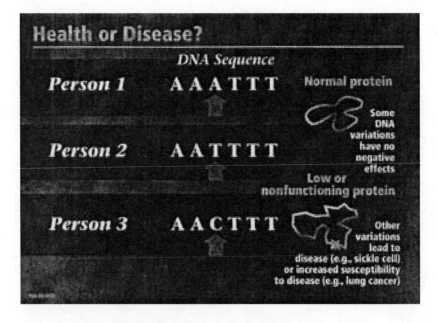

FIGURE 2.5. Genes and Disease
Used by permission of the U.S. Department of Energy Human Genome Program,
http://www.ornl.gov/hgmis.

folds back on itself to form a globule (tertiary structure). The structures of proteins determine the roles they play in organisms (Berman, Goodsell, and Bourne 2002). Proteins have many important biological functions. Some proteins are structural components of different parts of the cells and organisms. For example, muscle cell fibers, chromosomes, microtubules, mitochondria, and cell membranes all contain proteins. Thirty percent of all proteins are collagens (i.e., chemicals that are essential parts of bone, skin, hair, cartilage, and connective tissue). Other proteins play a key role in regulating cellular functions. For example, proteins can play a role in regulating DNA transcription and translation, cell division and growth, protein synthesis, cellular metabolism, and cell death. Some proteins, known as enzymes, play a key role in assisting important biochemical reactions. For example, hemoglobin is a protein that assists in oxygen transport. DNA-polymerase and RNA-polymerase are also enzymes. Other proteins are parts of hormones. For example, epinephrine, norepinephrine, estrogen, androgen, and glycogen are derived from proteins. Proteins are also part of biochemical markers known as antigens. The immune system is designed to respond to antigens that it does not recognize as belonging to the body (Walter 1982; Suzuki and Knudtson 1989; Kitcher 1997; Robertson 1994). Many proteins play a role in the active transport of ions, such as sodium and chloride, and in cell signaling and communication processes. Some proteins known as neurotransmitters play a crucial role in conveying nerve signals (action potentials) between nerve cells.

Proteins are a crucial link in the causal pathway from genotypes to phenotypes. For example, SCA is a phenotype that occurs when a person is unable to form a properly functioning hemoglobin protein as a result of genetic mutations. This disease is characterized by various signs and symptoms, such as sickle-shaped red blood cells, anemia, shortness of breath, and recurring episodes of severe pain (Walter 1982). Some phenotypes, known as monogenic traits, are caused by only one gene. When diseases, such as SCA, result from a single defective gene, they are known as monogenic diseases. Since an individual must lack both copies of a gene in order to develop SCA, the defective gene is said to be recessive, and SCA is therefore called an autosomal recessive disease (Resnik, Steinkraus, and Langer 1999). In some diseases, known as autosomal dominant diseases, it only takes one mutated allele to produce the phenotype. Although most of the genes that cause diseases are located on the autosomes, some occur on the sex chromosomes. For instance, hemophilia is caused by gene carried on the Y chromosome. Hemophilia is therefore a sex-linked disease. Finally, most of the diseases that have a genetic basis, such as cancer and diabetes, result from several different genetic defects. These diseases are known as polygenic diseases. Most of the phenotypes in human beings, such as height, weight, sensation, language, and cognition, are probably due to many different genes (Resnik, Steinkraus, and Langer 1999).

As one can see, the pathways from genotypes to phenotypes can be very complex and may involve interactions between many different genes and proteins. The newly emerging field of genomics aims to describe some of these interactions. Genomics seeks to describe and identify genes and explain their functions in organisms. Of course, biologists have known since Mendel's time that the environment also plays a key role in producing phenotypes. For example, someone with genes that predispose them to reach a full-grown height of six feet will not achieve that height if they lack an adequate diet. A person with genes that predisposes him to alcoholism may never develop this condition if he never drinks alcohol. Relatively few traits are causally determined by genes; the vast majority of phenotypes result from genetic, developmental, and environmental causes (Kaufman 1993; Robertson 1994). Rather than talking of the genetic causes of traits, it is often more appropriate to use terms like causal factors, predispositions, or risk factors (Kitcher 1997).

GENES AND MEDICINE

In mid 1980s, the U.S. Department of Energy (DOE) initiated a series of discussions about sequencing and mapping the entire human genome (Roberts 2001). Many scientists and politicians were skeptical about this enormous undertaking. Some argued that the money could be better spent on sequencing individual genes and studying their function instead of sequencing the whole genome, since the genome contains a great deal of junk DNA. The Human Genome Project (HGP) began in 1990, under the direction James Watson of the NHGRI. From 1990–2000, the United States contributed over $2 billion to the HGP, which was completed in February 2001, well ahead of schedule. One reason why the project was completed so early is that in the mid-1990s, private corporations, such as Celera Genomics also took an interest in the sequencing the human genome. Celera helped to accelerate the pace of research by raising hundreds of millions of dollars for genomics research, by sharing data with NHGRI, and by helping to transform the research environment from an academic endeavor to a high-stakes race to sequence the genome (Roberts 2001).

Technological developments in the last decade have also played a large role in accelerating the pace of recent research in genetics. Celera utilized both of these developments in its shotgun approach to sequencing the human and fruit fly genomes. In the 1990s, Perkin Elmer, a private company, developed automated sequencing machines and sold them to university and commercial laboratories. The machines can sequence thousands of DNA sequences in a single day. In that same decade, biologists and computer scientists started using powerful computers to process and analyze the vast sums of data involved in genetics, genomics, and proteomics research. In the twenty-first century, the new field of bioinformatics will play an essential role in helping researchers to understand the structures and functions of genes, proteins, and other macromolecules with biological significance (Roberts 2001).

One of the first big payoffs of the HGP has been a better understanding of the role of genetics in human health and disease (Collins and McKusick 2001). Genetic factors play a key role in many human diseases. About 5,000 mutations cause genetic diseases, if one understands cause to mean that it is practically certain that a person will develop the disease if he has the genotype for the disease. For instance, the Huntington's disease gene causes this disease because a person with one copy of this gene will almost certainly develop the disease. Geneticists also use the term penetrance to measure the percentage of people with a genotype who will develop the phenotype. A disease gene is 100 percent penetrant if every person with the gene will develop the disease. Huntington's disease is more than 95 percent penetrant (Resnik, Steinkraus, and Langer 1999). If we think of a genetic disease as a condition, such as SCA, which is caused by a heritable mutation, then nearly 10 percent of the human population has a genetic disease and each person carries about five recessive genes that can cause a disease (Kitcher 1997). Many more genes may predispose people to disease or increase a person's risk of developing a disease. For example, BRCA I and BRCA II mutations predispose people to develop breast cancer: women with these mutations have an 80 percent lifetime risk of developing breast cancer as compared to 10 percent for women who lack the mutation. When one includes genetic predispositions in the tally of genetic diseases, it is estimated that 60 percent of the human population has diseases with a significant genetic influence (Collins and McKusick 2001). However, many of the diseases that have a genetic influence also have many other causes. For instance, deafness can result from mutations as well as from damage to the inner ear from diseases or injuries. Deafness is highly variable because it has many different genetic and nongenetic causes. On the other hand, Down's syndrome has low variability because it has no known causes other than the mutation that produces three copies of chromosome 21.

Most of the key medical applications of genetic research will come through a better understanding of proteins, since genes contain information for making proteins, and proteins control many vital cell structures and functions (Collins and McKusick 2001; Human Genome Project 2001). In any particular metabolic process, there may be many different types of proteins involved as enzymes, as hormones, or as basic structures. In order to influence this process, one can develop drugs that target different biochemical pathways or reactions. If the process is a disease, one may develop drugs to treat the disease by developing a better understanding of the proteins involved in the disease. For example, in order to develop cancer treatments, one could use genetic information to develop drugs to block the formation of blood vessels that supply blood to tumors; one could use genetic information to develop drugs that

trigger aptosis (cell death) in cancer cells; one could develop drugs that acti-
vate inactive tumor suppressor genes; and so on. Since many antigens are
composed of proteins, researchers can also use genetic information to develop
vaccines for pathogens by synthesizing the pathogen's antigens. The vaccine
can be injected to induce an immune response. Because many hormones are
made out of proteins, proteins also hold the key to diseases that involve the
endocrine system, such hypothyroidism (Price and Wilson 1982).

Consider, for example, how genetic information on hypertension may
lead to drug therapies. It has been known for some time that hereditary factors play a role in hypertension, although researchers still do not understand
all the genes involved or the links between genetics and hypertension (Walter
1982). One recently discovered mutation responsible for early-onset hypertension suggests some ways of treating this disease as well as other forms of
the disease (Wicklegren 2000). The mutation causes malformation of the
mineralocorticoid receptor, a protein found in kidney cells, which regulates
salt retention. A normal mineralocorticoid receptor causes kidney cells to
absorb salt in response to the hormone aldosterone. Salt absorption leads to
water retention, which raises blood pressure. The mutated receptor, however,
always causes the kidneys to absorb salt. Although this is a rare form of hypertension, the research suggests that similar chemical pathways may be involved
in other forms of hypertension. It may one day be possible to treat hypertension by developing drugs that inhibit hormones involved in hypertension or
by blocking or modifying mineralocorticoid receptors. Some of these drugs
may themselves be proteins. In any case, it will be important to be able to
manufacture the proteins involved in hypertension to study how they affect
and regulate sodium absorption.

For many years, pharmaceutical companies have developed drugs that
have some type of action on proteins, are composed of proteins, or contain
proteins. The drugs often work by inhibiting, buffering, or catalyzing chemical reactions in the body. For example, insulin, which is used to treat diabetes,
is a hormone that catalyzes the conversion of glucose into glycogen in the
body. Furosemide, which is used to treat edema, affects the chloride transport
systems in kidneys by blocking the reuptake of chloride, thereby causing the
body to excrete water. Many of the drugs that have psychopharmacological
effects work by accelerating the transmission of neurotransmitters, blocking
the reuptake of neurotransmitters, or decreasing or increasing the number of
neurotransmitter receptors.

Identifying, finding, isolating, and purifying proteins in the body can be
arduous and tedious processes, since the 2 million different proteins in the
human body are spread throughout the whole body and are manufactured at
different times. In the past, drug discovery has proceeded largely by trial and
error as scientists attempted to develop compounds that affect proteins in the
body. Genetics can promote and augment drug discovery by providing

researchers with readily available information about the primary structures of proteins. With this knowledge, it may be possible to design drugs that can affect these proteins. Instead of proceeding by pure trial and error, it will be possible to use genetic information to tailor specific treatments to specific problems in the body. The new field of proteomics aims to identify, isolate, and purify proteins and describe their interactions and functions (Fields 2001).

Thus, genes have not only scientific but also medical and economic value because they are the key to drug discovery and development in the twenty-first century (Subramanian et al. 2001). In addition to using genetic information to develop new drugs, it may also be possible to use genetic information to get a better understanding of how individuals respond differently to the same drugs. Pharmacists, physicians, and pharmacologists have known for some time that patients respond differently to the same medication. One person may tolerate an antibiotic, while another may have a severe reaction; one person may derive great benefits from a particular analgesic, while another may not. Since differential responses to drugs probably result from metabolic variation, and metabolic variation is probably due, in part, to genetic variation, researchers and pharmaceutical companies may be able to use genetic information to design drugs, develop drug combinations, or recommend dosages tailored to individuals or genotypes. Thus, the genetic revolution has given rise to a new biomedical field known as pharmacogenomics (Wickelgren 1999; Service 2000).

In addition to its impact on drug development and pharmacogenomics, genetics has already had significant applications in genetic testing (Collins and McKusick 2001). Researchers have developed tests for a variety of genetic conditions including breast cancer, heart disease, Alzheimer's disease, CF, SCA, Huntington's disease, and Canavan's disease. Although many of these diseases, such as Huntington's disease, are currently not treatable, information from genetic tests can be useful to prospective parents or adults who want to use the information to prevent diseases or make life plans (Kitcher 1997). For example, if a couple both test positive for the CF allele, they will know that they have a 25 percent chance of having a baby with CF, and they may decide to use a procedure, such as preimplantation genetic diagnosis, in order to give birth to a healthy child (Resnik, Steinkraus, and Langer 1999). If a woman tests positive for BRCA1 or BRCA2 mutations, then she may decide to take tamoxofin to lower her chances of developing breast cancer, she may make some changes in her diet, or she may decide to have a prophylactic mastectomy.

As a result of advances in DNA chip technology, it is now possible to test for thousands of different genes at one time with a single assay. A DNA chip contains thousands of different strands of single-nucleotide DNAs known as DNA probes. These single nucleotides can bind with (or hybridize with) their mirror image strands in the test sample. Since researchers know the sequences of DNA in the strands used in the assay, they can infer the sequences in the sample from the areas on the chip where hybridization occurs. DNA chips that have been developed thus far can test for up to 400,000 unique DNA strands. The data from DNA chips can also be used to study patterns of SNPs and haplotypes among individuals and to analyze gene expression in tumors and diseased tissues (Maughan, Lewis, and Smith 2001). Advances in genetic testing raise many important issues relating to the confidentiality of genetic information, genetic discrimination, and eugenics, which I will not explore here (see Kitcher 1997; Rothstein 1997; Mehlman and Botkin 1998; Resnik, Steinkraus and Langer 1999).

The field of gene therapy uses the gene transfer techniques (discussion to follow) to replace defective genes. As this new field continues to progress, it may be possible to treat many diseases by replacing defective genes, transferring normal genes into to the body, or by transferring genes to specific targets in order to trigger immune responses (Collins and McKusick 2001). For instance, if gene therapy for SCA ever becomes a clinical reality, it may be possible to deliver normal hemoglobin genes to SCA patients so they can manufacture hemoglobin. One day it may be possible to use gene therapy to deliver genes to cancer cells that would trigger aptosis or would make the cells express an antigen that would cause the body's lymphocytes to start destroying the cancer cells. Gene therapy raises a number of ethical issues, such as the appropriate protections for human subjects and eugenics, which I will not address in this book (see Walters and Palmer 1997; Resnik, Steinkraus, and Langer 1999). However, I will consider some of the intellectual property issues related to gene therapy.

Finally, it should be mentioned that there are many possible medical applications of genetics, which have not been clinically tested but may one day have a profound importance in medicine, such as stem cell therapy, therapeutic cloning, and organ and tissue engineering. Genetic information is likely to play a key role in these emerging fields because it will enable researchers to modify, develop, or design cell lines, tissues, or organs with specific genetic characteristics (Juengst and Fossell 2000). These new technologies will probably employ gene transfer techniques (discussion to follow) to allow researchers to genetically engineer human cells, tissues, and organs to overcome tissue rejection problems.

GENETIC ENGINEERING AND BIOTECHNOLOGY

Although DNA replication and protein synthesis are natural processes, scientists can manipulate these processes to reproduce DNA and proteins under laboratory conditions. Since the early 1970s, molecular biologists have been able to use bacteria to clone or modify DNA and proteins. Although bacteria do not use sex to share DNA, they are able to exchange DNA through plasmids, which are vesicles that contain DNA. The exchange occurs when one bacterium emits a plasmid vector that binds to the surface of another bacterium. A vector is any type of structure or mechanism for transferring DNA across cells or species. Other vectors include viruses and artificial chromosomes. It is possible to transfer DNA without vectors, but vector-aided transfer is usually more efficient than vectorless transfer (Resnik, Steinkraus, and Langer 1999). After the vector binds to the membrane of the bacteria, the DNA penetrates the membrane and is incorporated into the host's genome, which consists of the DNA in bacteria's single circular chromosome as well as the DNA from plasmids. When the bacteria reproduce, they produce a copy (or clone) of the DNA they receive. If the DNA codes for a protein, the bacteria also produces a copy (or clone) of that protein. This kind of DNA exchange has taken place for hundreds of millions of years and plays a key role in bacterial evolution. Bacteria can evolve very rapidly because they reproduce quickly and are able to exchange genes that may confer an adaptive advantage under particular environmental conditions. For example, if a bacterium develops resistance to an antibiotic, the gene for this resistance can be transferred to other bacteria (Suzuki and Knudston 1989).

Modern cloning techniques make use of these natural processes. The cloning of DNA involves basically the same steps. First, scientists treat a cell with chemical and mechanical processes to extract its DNA. Then they use enzymes to chop the DNA into its component parts. After breaking down the DNA, scientists can use chemical and electrical processes to sort the DNA into parts of different lengths. Next, they may select one of these segments to be incorporated in a plasmid vector. To incorporate the sequence into a plasmid, scientists use restriction enzymes to splice the sequence into the plasmid's DNA ring. They then allow the vector to transfer the DNA to the bacteria, usually a strain of *E. coli*, which incorporate the DNA into their genomes. To obtain a purified sample of the DNA, scientists can apply the same extraction techniques used on the cell sample to the bacteria to extract its DNA. To make large quantities of the cloned DNA, researchers can allow the bacteria to reproduce (and therefore manufacture more DNA) or they can use the polymerase chain reaction (PCR) to manufacture more of the DNA under laboratory conditions (Wong 1997; Lucotte and Baneyx 1993). In either case, researchers end up with a sample of isolated and purified DNA sufficient for conducting additional research and product development (see table 2.1).

In addition to cloning DNA sequences, researchers can also use natural processes to modify those sequences. For example, after extracting a DNA sequence from a cell, researchers could treat the sequence with enzymes that cause changes in the DNA, such as deletion of DNA sequences, addition of DNA sequences, replacement of DNA sequence, or transposition of DNA

TABLE 2.1
Isolating and Purifying DNA

Step	Process
1	Obtain a cell sample
2	Use chemical and mechanical processes to break down cells into their chemical components
3	Use a centrifuge to separate the DNA from other chemicals
4	Use restriction enzymes to cleave the DNA to produce pieces of DNA (genes and gene fragments)
5	Use gel electrophoresis to separate the selected DNA from the other pieces of DNA
6	Use restriction enzymes to splice the selected DNA into plasmids
7	Insert plasmids into bacteria
8	Allow the DNA from the plasmids to recombine with bacterial DNA
9	Culture bacteria
10	Identify the bacteria that reproduce the selected DNA
11	Culture the bacteria that reproduce the selected DNA
12	Repeat Steps 2–5 to separate the selected DNA from other pieces of bacterial DNA
13	Use PCR to produce a larger sample of the selected DNA

sequences. The resulting sequences can then be transferred, cloned, isolated, and purified. Researchers can thus produce sequences of DNA that are not the same as the sequences found in nature. The degree of similarity between two macromolecules is known as the percent homology: sequences that are 100 percent homologous are identical, while sequences that are 95 percent homologous are very similar. The methods of modifying DNA can also by applied to protein synthesis to yield modified proteins. By modifying the DNA that is transferred to the bacterium, researchers can produce proteins with different primary structures. Thus, a new protein might be 95 percent homologous to a protein occurring in nature. The concept of homology also applies to the relationship between a DNA fragment and a larger sequence. A fragment might have a 10 percent or less homology to a larger sequence. Some important DNA fragments, know as expressed sequence tags (ESTs) can be used to locate a sequence that codes for a protein. However, ESTs might contain 10 percent or less of the DNA found in the entire coding sequence (Fox and Paul 1999).

The gene transfer techniques discussed above have been used to genetically engineer cells, tissues, and organisms. Bacteria were the first organisms to be genetically engineered. In the next chapter, we will discuss an important case in patent law, *Diamond v Chakrabarty* (1980), which involved genetically engineered bacteria. For now, we shall briefly consider some of the applications of genetic engineering in agriculture. One early application of genetic engineering was to use bacteria to manufacture human hormones, such as growth hormone. It is very difficult to obtain human growth hormone in large quantities from people, since it must be extracted from the pituitary gland, which is located in the center of the brain. Biotechnology can design bacteria that make human growth hormone by taking a human cell, extracting its DNA, finding the gene that codes for growth hormone, inserting that sequence into a vector, and allowing the vector to transfer the DNA to the bacteria, which expresses the gene as human growth hormone.

Gene transfer techniques can also be used to transfer genes into livestock or crops for agricultural and medical purposes. For example, the long-term goal of cloning sheep is to be able to genetically engineer sheep that produce a variety of human hormones in their milk, such as factor IX, a blood-clotting protein (Wilmut 1998). It may be possible to use sheep and many other animals as drug factories some day. Another possible application of recombinant DNA is to transfer human genes into pigs to enable them to serve as organ donors for human beings. Pigs are anatomically similar to humans, but there are several current barriers to xenotransplantation (transplantation across species), including the possibility of zoonoses (diseases transferred across species) and tissue rejection. Genetic engineering may help overcome the tissue rejection problem by enabling researchers to create pigs that express human histocompatibility proteins in their tissues and organs (Chapman and Bloom 2001).

Researchers have also used genetic engineering techniques to develop animals for use in research. In the late 1980s, Harvard University and Dupont developed a genetically engineered mouse that was designed to develop specific types of cancer. The mouse, known as an oncomouse, was created with a propensity to develop certain forms of cancer and provided researchers with a useful animal model for cancer treatments. Harvard and Dupont were issued a patent on the mouse on April 12, 1988 (Kevles and Berkowitz 2001). Since then, researchers and biotechnology companies have developed genetically engineered mice designed to model a variety of human diseases, including diabetes, multiple sclerosis, Alzheimer's dementia, and obesity. The laboratory mice business is now a $200 million a year industry and many of those mice are genetically engineered varieties. Researchers can alter the mouse genome by either deleting (or knocking out) a gene or inserting a gene. In both cases, it is possible to study the biochemical and physiological effects of the gene in the mouse in order to understand the pathways of disease and possible interventions (Malakoff 2000). In 2001, researchers used similar procedures to create a rhesus monkey known as ANDi. In the procedure that created ANDi, scientists stitched a jellyfish phosphorescence gene into a viral vector and infected rhesus monkey eggs with the virus. ANDi was born with a functioning phosphorescence transgene (a transferred gene), which was verified by the fact that ANDi's tissues exhibited phosphorescence (Chan et al. 2001). The long-term goal of this researcher program is to develop genetically engineered primate models for human diseases.

Finally, it is important to mention that for over two decades researchers have applied recombinant DNA techniques to produce genetically modified (GM) plants, crops, and foods. In some case instances, researchers have transferred genes into plants in order to protect them from shelf rot and cold damage. For instance, scientists have developed GM tomatoes that contain fish antifreeze transgenes. These fish genes allow the vegetables to have extra tolerance for refrigeration. Others have transferred genes into plants in order to increase crop yield, to boost drought tolerance, to enhance nutritional value, or to provide resistance to herbicides and pesticides. Some researchers have developed GM grass that would resist drought and require little mowing. Environmentalists have raised concerns about the risks of GM crops to the environment, while public health advocates have expressed concerns relating to public health and safety. Other critics have objected to licensing and marketing practices used by biotechnology companies that commercialize GM crops and plants. GM foods, plants and crops have raised a variety of social, political, and ethical issues, which we will consider in Chapter 9 (Enserink 1999).

SUMMARY

As one can see from this brief survey, research and product development related to DNA plays a key role in basic sciences, such as genetics and genomics, as well as in applied disciplines, such as medicine and agriculture. The ability to discern, replicate, manipulate, interpret, and analyze information contained in the genomes of humans, animals, and plants is the cornerstone of biomedical and biotechnological R & D in the twenty-first century. This information has technological, scientific, and political significance and therefore also has a great deal of economic value. Someone who invents a new assay, procedure, product, or technique related to the use of DNA in biomedicine and biotechnology could earn a great deal of money. In the last decade, individuals, private companies, and academic institutions have asserted intellectual property rights over DNA as they have tried to stake their claims in this genetic gold rush. Should society honor these rights? Should DNA be privately owned? The remainder of this book will address these and other moral issues, but first we must examine the legal context for these debates.

3

DNA as Intellectual Property

WHAT IS INTELLECTUAL PROPERTY?

All properties can be viewed as a collection of rights to control something (Honore 1977; Benn 1967; Dreyfuss 1989).[1] A person who has property rights with respect to a plot of land, has rights to sell, modify, possess, use, mine, and rent the land. Although we tend to think of the land as the property, the land is a mere thing that has its status as property in relation to members of society. If no people were around to use language to ascribe property rights to the land, the land would not be someone's property; it would just be the land. The deed to the land entitles the person to these property rights pertaining to the object, the land. The deed, like other legal documents, is a social recognition of the person's rights and duties. This is a very important point to remember about property, because our natural tendency is to objectify property and endow it with some independent existence. We must remember that property is essentially a social institution that serves various purposes in society (Feinman 2000).[2]

Societies have developed a variety of forms of proprietary control over objects. The most extensive form of proprietary control is what legal scholar Margaret Radin calls complete commodification (Radin 1996). Commodification (or commercialization) is a process in which societies apply market rhetoric, concepts, and values to objects. If something is completely commodified, it is a complete commodity; if something is only partially commodified, it is an incomplete commodity; if something is not commodified at all, then it is a noncommodity. Complete commodities are: alienable, that is their control can be transferred from one person to another; commensurable, that is

31

they can be compared to other objects in terms of a common currency, such as money; and fungible, that is they can be replaced by other things of the same kind with no loss of value (*Black's Law Dictionary* 1999). Objects that are treated as complete commodities, such as corn, oil, or steel, can be bought, sold, manipulated, possessed, used, destroyed, or rented with very little government regulation. Ownership, on this view, is a form of proprietary control that implies a variety of property rights, such as the right to possess, use, and convey some thing (*Black's Law Dictionary* 1999). Ownership may involve complete commodification, such as when one owns a bushel of corn, or incomplete commodification, such as when one owns a pet.

Incomplete commodification occurs when societies decide to regulate the commercialization of objects in order to preserve other (nonmarket) values. For example, dogs and cats are incomplete commodities. Although society allows pets to be bought and sold, trained and bred, society also has laws that forbid inhumane treatment of animals and require vaccinations as a condition of ownership. Other incomplete commodities include houses, cars, health care, guns, phone service, gambling, professional sports, electricity, human labor, and alcohol. Although the U.S. operates under a free market system, the nation regulates many of its economically and socially significant goods and services in order to protect individual rights, to prevent harm to individuals, or to promote government interests (Radin 1996).

Finally, there are some things that society does not treat as commodities at all, such as voting rights and legal rights, and, according to many, love. The reason why some things are kept off the market is that people believe that these things will be destroyed by the incursion of any market values. Allowing people to buy votes undermines democracy, and love that can be bought or sold is not love at all. A key policy issue with regard to commodification, which we will explore in chapter 6, is the question of whether there is a slippery slope from incomplete to complete commodification (Radin 1996). By treating something as an incomplete commodity, do we run the risk of one day treating it as a complete commodity as a result of changing attitudes?

Most people think of their property in very concrete terms, such as my doll, my car, my house, or my land. These types of property are known as tangible properties because the things that are treated as property occupy a particular place in space at a particular sequence in time (Miller and Davis 2000). Intellectual properties, such as copyrights and patents, are intangible properties because they convey property rights over abstract things that do not have a specific location in space or duration in time (*Black's Law Dictionary* 1999). A computer program, a song, a poem, and an invention have no particular location in space or duration in time. Not all intangible properties are intellectual properties: some intangible properties, such as stocks, bonds, annuities, pensions, are financial instruments. Intellectual properties are called intellectual because they have to do with concepts and ideas, not with money. (But, as we shall see, this difference is not as large as one might think.)

Intellectual property (IP) is also what economists call a nonrivalrous good, because it can be shared without reducing the owners' ability to use it (Dreyfuss 1989). For example, two people cannot both use the same toothbrush or baseball glove at the same time, but they can both use the same song, computer program, name, or chemical formula at the same time. It is easy to understand why societies would enact laws to protect rivalrous goods, since these laws help settle claims about physical possession and control: if I own a plot of land, then I have the right to occupy it and to keep you from using it. If you use my land, then you are diminishing my ability to use it. But these types of disputes need not arise with respect to intellectual property, since two people can both use the same idea or concept without interfering with each other's ability to use it. If intellectual properties can be shared without reducing our ability to use them, then why would societies enact laws in order to give people rights to control these properties? Why shouldn't all ideas be shared freely, without any legal restrictions? To answer these questions we need to examine moral and political justifications for intellectual property (Hettinger 1989). In order to understand these justifications it will be useful to examine their historical context first.

HISTORICAL ORIGINS OF INTELLECTUAL PROPERTY

The world's first patents were issued in England in the 1400s when the Monarchy granted privileges, known as Letters Patent, to manufacturers and traders. In 1449, King Henry VI issued the first known English patent to John of Utynam for a method of making stained glass (U.K. Patent Office 2001). In exchange for these privileges, the manufacturers and traders were required to teach their art to others. During this time, many manufacturers, traders, and others attempted to retain monopoly control of their businesses through trade secrecy. From the very beginning, patents were viewed as a bargain between the government and private citizens for the public disclosure of information. In 1623, England passed the Statue of Monopolies to end some of the disruptive monopolies that were affecting commerce and industry (Miller and Davis 2000).

During the next two hundred years, patents became a routine part of commerce and industry in England. In 1765, the British government awarded a patent to James Watt (1736–1819) titled "A New Method of Lessening the Consumption of Steam and Fuel in Fire Engines." Watt's steam engine helped to provide additional justification for patents and served as a model of science-industry collaboration. Watt collaborated with entrepreneurs John Roebuck and Matthew Boulton. Roebuck, who made two-thirds of the initial investment in the engine, went bankrupt. Boulton bought Roebuck's share of the patent and helped to market this new product. Watt and Boulton made a considerable sum from the steam engine, which was the product of scientific ingenuity and private investment and marketing (Burke 1995).

In 1641, the Massachusetts Bay Colony awarded the first Colonial patent for the production of salt. After the Revolutionary War, the framers of the U.S. Constitution were keenly aware of the connections between science, technology, industry, and the practical arts. Thomas Jefferson (1743–1826) was himself an author and inventor, and Benjamin Franklin (1706–1790) was both a statesman and a prolific inventor. The utilitarian rationale for IP can be found in the U.S. Constitution (1787), Article 1, Section 8, Clause 8, which states that "The Congress shall have the power . . . to promote the progress of science and the useful arts by securing to authors and inventors the exclusive right to their respective writings and discoveries." In 1790, the U.S. Congress enacted the first patent laws, which have been amended numerous times, including significant revisions in 1952, 1984, and most recently in 1995. The Patent and Trademark Office (PTO), which was established in 1836, administrates, oversees, and awards patents in the United States (Patent and Trademark Office 2001; U.S. Patent Act 1995).

One can find a similar tension between public and private control of IP in the history of copyright law in the United States and England. Before the invention of the printing press in the 1500s, people copied books and other documents laboriously, by hand. Since it took so much effort to copy a book, the problem of unauthorized copying of books did not arise very frequently. After the printing press was invented, it was possible to make thousands of copies with relative ease. Soon companies sought to control the publishing industry in the same way that manufacturers and traders had developed monopolistic control over commerce (Miller and Davis 2000). Authors and others challenged the power of publishers and demanded some of their own rights and privileges. Thus, the idea of a copyright was developed in eighteenth century England as a way of helping authors and publishers to settle these sorts of questions. In 1710, the English Parliament passed a statute granting copyright protection to books and other writings. Prior to this statute, copyrights were protected only by common law. In 1790, the United States enacted the Copyright Act, which was revised in 1831, 1870, 1909, and 1976 (Miller and Davis 2000). As we shall see below, copyright protections have been extended from books to many other original works.

MORAL FOUNDATIONS OF INTELLECTUAL PROPERTY

As one can see from this brief sketch of the history of IP, two major theories have been invoked to justify IP: utilitarianism and libertarianism (or the natural rights approach) (Miller and Davis 2000). The most influential justification for intellectual property appeals to some type of utilitarian moral theory. According to a version of utilitarianism known as rule utilitarianism, policies and rules should be evaluated with respect to their likely effects (both good and bad) on society (Scheffler 1988). The best policy is the one that is likely to produce the most net social utility (i.e., maximum benefits and minimum harms) (Mill [1861] 1979). According to utilitarians, laws and policies that establish and regulate intellectual property should be judged according to their likely effect on social goods, such as human happiness, health, scientific progress, or economic development. According to those who take this approach to IP, society should recognize certain forms of intellectual property because this will promote socially valuable consequences, such as scientific discovery, technological innovation, artistic creativity, medical progress, entrepreneurial activity, business investment, or economic growth (Kuflick 1989; Hettinger 1989; Svatos 1996; *Kiwanee Oil Co. v Bircon Corp.* 1974).

IP laws encourage the progress of science and the useful arts in several ways. First, they provide authors and inventors with economic incentives to share the products of their labor with the public. Without such protections, authors and inventors might decide not to pursue their original works or inventions or to keep them a secret (Foster and Shook 1993). Second, IP laws encourage businesses (and entrepreneurs) to invest in research and development (R & D) by enabling them to secure reasonable returns on investments. Without such protections, businesses would either not invest in R & D or would use trade secrecy to protect their proprietary interests related to R & D. Third, IP laws help provide a stable regulatory climate for companies to conduct business and market products. Economic development depends, in part, on a stable system of laws that regulate trade and commerce (Samuelson 1980).

The other most influential justification for IP rights appeals to libertarian ideals relating to property. This approach to IP traces its history to the seventeenth century British political philosopher John Locke (1632–1704) ([1764] 1980), who founded the natural rights approach to politics and morality. According to this view, all people have inherent rights to life, liberty of thought, liberty of action, liberty of expression, and property. The sole reason for the existence of government is to protect these rights and the government's authority extends no further than this mandate permits. Thus, the only legitimate reason to restrict an individual's rights is to prevent him or her from violating someone else's rights (Nozick 1974). Since rights protect individual interests, and harms can be understood as actions that adversely affect interests, most natural rights theorists accept the harm principle: a person's rights may be restricted in order to prevent him or her from harming other people (Feinberg 1973). For example, speech, in general, should be free but the state may restrict specific types of harmful speech, such as slander or libel. The natural rights approach is a type of deontological theory because it holds that people have inherent rights that do not depend on consequences (Frankena 1973).

According to Locke's approach to property, there are some natural resources that initially belong to no one, such as air, water, land, plants, animals, and so on. People own their bodies and they may appropriate things from nature by adding their labor to those things. People cannot own things that cannot be modified or created by human labor, such as the sun. Human labor transforms common resources into private resources and also creates new resources from existing ones. For example, a person may come to own wild donkeys by using her labor to capture those donkeys. Although she owns this resource, she has not made anything new yet. However, if she uses the donkeys to plow a field and plant and harvest a crop of corn, then she has made something new from natural resources. By mixing labor with something, a person transforms that thing in such a way that it is fundamentally connected to that person. For example, a flute carved from a piece of wood reflects the skill and artistry of the flute maker and is therefore his property.

People may also own property that they acquire through barter, exchange, or as a gift, provided they acquire the property without violating someone else's rights. Fraud, theft, and exploitation are illicit ways of acquiring property (Nozick 1974). The labor mixing theory can also be applied to intellectual property as well; a person may own an original work, an item of information, a name, or an invention insofar as the thing is a product of his or her labor. An invention, song, or book is mine because I put my labor into it and I deserve to benefit from it (Kuflik 1989). Someone who uses, copies, or sells that invention, song, or book, without my permission, violates my rights to the products of my labor.

An approach to IP that also supports a libertarian perspective draws on the work of the nineteenth century German political philosopher Friedrich Hegel (1770–1831). Several political philosophers and legal theorists, such as Radin (1996) and Waldron (1988), have applied Hegel's ideas about property to contemporary controversies. Hegel (1976) held that property provides a medium for the development and expression of one's self. All people have a basic need to define themselves in relation to the world, which is separate from the self. One defines the self by manipulating and controlling objects in a medium separate from the self. In order to ensure that people can have the degree of control over objects in the world required for self-expression, society should recognize property rights. Since a person cannot be truly free unless she is able to express herself in the world and develop a life plan, property rights are morally justified because they promote human freedom (Waldron 1988). Limitations on property rights, including IP rights, therefore involve restrictions on human freedom.

Finally, one cannot discuss property rights without mentioning the views of Karl Marx (1818–1883), a German political philosopher and economist, who studied Hegel but who reached very different conclusions about property. Marx's theory of history, known as dialectical materialism, helped to inspire communist revolutions in the former Soviet Union and the People's Republic of China. According to Marxists, all private property should be abolished because property alienates workers from their labor and allows the bourgeois class to exploit the proletariat class. In a capitalistic society, labor is converted into capital, which can be traded on the market. The bourgeois exploit the proletariat by acquiring and accumulating capital. In a pure socialistic society, there is no private property, including intellectual property. All property belongs to the state, which means, in effect, that individuals do not have any property rights since the state reserves the right to control property (Marx [1867] 1996; Marx and Engels [1848] 1998). I mention the Marxist view here as an alternative to utilitarianism and libertarianism, which both provide a rationale for IP. Although I do not favor a Marxist approach to IP, I think the theory is worth considering because it articulates the idea of common property or a public domain. Even someone who accepts the standard justifications for IP might recognize the need for some types of common property. We will return to this idea many times in this book as we explore ways of striking a balance between public and private control of IP.

RESTRICTIONS ON IP RIGHTS

So far we have discussed approaches to justifying IP but we have not addressed arguments for limiting or restricting IP rights. Consider what would happen if people had unlimited IP rights. What if people could patent the sun, water, breathing, algebra, or writing? What if people could copyright the letters of the alphabet, the words in the English language, walking, or basic human sounds? Obviously, excessive IP rights can have disastrous social consequences. Even a libertarian would admit that IP rights may therefore be restricted to prevent harm to individuals. Utilitarians would go further and say that these rights can be restricted in order to prevent harm to society and to promote social goods.

If we think of this issue in more general terms, restrictions on IP rights can be understood like other restrictions on rights. There are two well-recognized ethical and policy arguments for imposing some restrictions on individual rights. The first reason, known as the private harm principle, is that individual rights may be restricted in order to prevent harm to others (Mill [1859] 1956). For example, no one has a right to kill, assault, batter, or rape someone else. One does not even have a right to impose a significant risk of harm on others by shouting "fire" in a crowded theater or by driving recklessly (Feinberg 1973). Unrestricted IP rights often result in a failure to benefit other people, but failing to benefit a person whom one has no special relationship with, such as doctor-patient or parent-child, is not the same thing as harming a person.

A second reason for restricting individual rights, known as the public harm principle, holds that rights may be restricted in order to prevent individuals from causing harm to practices or things that serve the public interest (Feinberg 1973). For example, this principle would imply that one does not have the right to litter, to disrupt transportation, or to interfere with voting, since the public has an interest in having a clean environment, convenient transportation, and unfettered voting. Since failing to benefit a public interest is not the same thing as harming a public interest, there is a third principle, known as the social welfare principle, which holds that rights may be restricted in order to benefit others or promote public interests. For example, this principle might be used to justify requiring people to register for the draft or send their children to school.

In examining restrictions on rights, we must balance several different considerations, including (Feinberg 1973):

1. The overall social value of the rights in question.
2. The burdens imposed by restricting rights.
3. The overall probable harms and benefits of restrictions on rights.

For example, consider restrictions on smoking in public. In debating about a law that bans smoking in a public place, society should address the overall social value of the right to smoke in public, the burdens on smokers who are not allowed to smoke in public, and the benefits and harms of restricting the right to smoke in public, including benefits from curtailing exposure to second-hand smoke as well as possible harms to smokers and certain businesses. In making these types of policy decisions, one must consider and weigh the different values at stake as well as the relevant facts, since many of the facts, such as the facts about the effects of second hand smoke, play a key role in how we should balance the various values that come into play.

The same type of analysis applies to restrictions on IP rights. In deciding whether to restrict IP rights, we need to balance the social values at stake (e.g., individual rights versus the common good). We also need to consider the relevant economic, scientific, technical, and cultural facts, since these have an important bearing on how we should balance the different values at stake. Since most people will agree that there should be some IP rights but that these rights should not be absolute, the most basic ethical and policy issue we face in the area of intellectual property is how to provide the appropriate balance of public versus private ownership and control. What type of information should be in the public domain and what type of information should be under private control? Almost all IP controversies reflect different perspectives on this fundamental question (Dreyfuss and Kwal 1996; Hettinger 1989). As we shall see below, IP laws and policies all reflect particular approaches to this question.

PATENTS

This book will focus on patent laws in the US, which are similar to the patent laws in Europe. The book will address European patent laws insofar as they pertain to the issue of patenting DNA. Under U.S. law, patent rights granted by the PTO give inventor(s) the right to exclude others from producing, using, or commercializing their inventions for twenty years from the date that the application is submitted. The emphasis here should be on excluding others, since patents do not actually give inventors any positive rights to make, use, or commercialize their inventions. For example, the fact that I can patent a radar detector gives me no right to use this invention in a particular state if it is illegal to use that invention in that state. I also do not have a right to make or use an invention if it infringes on someone else's patent. Double-patenting, patenting the same invention twice, is not allowed under U.S. law (Miller and Davis 2000). (However, companies have tried to find ways around this law by patenting new applications of their invention or similar inventions).

In order to obtain a patent, inventors must submit an application to the PTO. Patentable items include "processes, machines, articles of manufacture, compositions of matter, or new improvements thereof" (U.S. Patent Act, 1995, sec. 101). The definition essentially divides patents into three types: (1) patents on process, (2) patents on products, and (3) patents on improvements (Miller and Davis 2000). For example, a patent on a light bulb would be a patent on a product; a patent on a method for manufacturing light bulbs would be a patent on a process. (Inventors often patent both the object and the process for making the object.) A patent on an improvement on a light bulb could include a patent on an improved design of the invention or a new application of the invention (e.g., to cook hot dogs). Although it is important to distinguish between these three types of patents, the bottom line as far as patent protection is concerned is that the invention must qualify under at least one of three types of protection (Miller and Davis 2000).

In the United States and in Europe, an item must be a product of human ingenuity to qualify as a patentable subject matter. U.S. courts have ruled laws of nature, natural phenomena, and naturally occurring animals, plants, and chemical compounds are not patentable (*Parker v Flook* 1978). The courts have interpreted the concept of a product of human ingenuity rather expansively in recent years, ruling that anything made by man is patentable (*Diamond v Chakrabarty* 1980). To obtain a patent under U.S. law (or European law), the applicant must provide evidence on his application that the invention meets the following conditions (Miller and Davis 2000; Nuffield Council on Bioethics 2002):

1. Novelty. The invention is new or original. It has not been previously invented nor has it been previously disclosed in the prior art, which includes publications, patent applications, and prior uses of inventions. The disclosure need not occur in a single act of invention, publication, or use in the prior art; disclosure can occur if the whole invention is disclosed in a series of different publications or inventions. Thus, the whole combination of different elements of an invention may constitute an enabling prior disclosure that makes the invention unpatenable (*In re Wiggins* 1973).

2. Nonobviousness. The invention is not obvious to a person trained in the relevant prior art. To decide whether an invention is nonobvious, the PTO must compare the invention to the prior art to determine whether those with ordinary skill in the prior art would have a reasonable expectation of success at making the claimed invention. (*Graham v John Deere Co.* 1966).

3. No statutory bars. The invention is not barred by statute for lack of diligence. The United States has laws that bar a person from obtaining a patent if they do not pursue their invention diligently by reducing it to practice and filing an application on time. In the United States, if two people both claim to have conceived of the same invention, the PTO will award the patent to the first person to conceive the invention, unless the first inventor does not diligently pursue the invention by reducing it to practice and submitting an application on time.

4. Usefulness. The invention has some specific, definite, practical use. The invention must have more than a hypothetical, open-ended or throwaway use. A patent is not a hunting license (*Brenner v Manson* 1966). The PTO does not presume that the invention is useful; the inventor must prove that it has a use. Proof of the potential use of the invention must be based on previous research and analysis, not on mere speculation. As noted in chapter 1, the PTO recently clarified its policy concerning the proof of utility for DNA patents (Patent and Trademark Office 1999).

In addition to showing that his invention meets conditions 1 through 4, the inventor must satisfy another condition on his application:

5. Enabling description: The invention is described in enough detail to allow others skilled in the art to make and use the invention. Although an inventor need not produce the actual invention or a scale model, he or she must describe the invention. In DNA patenting, U.S. courts now require inventors to provide a precise description of the patented DNA sequence (*Regents of the University of California v Eli Lilly and Co.* 1997).

Once the patent is approved, the application becomes a public document. Patent applications are available on the PTO's website (www.uspto.gov). Since the application becomes part of the public domain, patents involve a type of bargain in which a private right is traded for public disclosure (Miller and Davis 2000). As noted earlier, this bargain reflects some kind of balancing between public and private control over IP. By awarding this private right, the government gets something in return (public disclosure) and provides inventors and entrepreneurs with incentives. The social benefits of the public disclosure of scientific and technical information through patent applications are well documented. The were over 30 million patents in the world in the year 2000, with one million new patents added each year. Patents constitute the single largest body of technical information in the world (Derwent 2001a).

Without patent protection, many inventors would resort to trade secrecy. A trade secret is information with economic value that a business keeps confidential to maintain a competitive advantage over other businesses (*Black's Law Dictionary* 1999). To obtain trade secrecy protection, a business must take reasonable steps to maintain secrecy. State and federal laws in the United States protect trade secrets, and violators are subject to a fine, imprisonment, or both (Foster and Shook 1993). For example, the formula for Coca Cola is one of the most carefully guarded trade secrets. Although it is illegal to steal or improperly disclose trade secrets, it is legal to discover trade secrets by lawful means, such as independent research or reverse engineering. If Pepsi Cola wanted to obtain the formula for Coca Cola, the company could do its own research to try and discover the exact formula for the product and it could try to recreate the invention. One problem with trade secrets is that they are often short-lived and difficult to protect. Thus, many inventors and businesses prefer patenting to trade secrecy (Foster and Shook 1993).

Patent rights may be assigned or licensed. For example, a privately funded researcher may assign his patent to a company that funds her research in exchange for a fee or royalties. The company would then have exclusive rights on the invention. Patent holders may allow others to use, make, or commercialize their inventions by licensing their inventions in return for a percentage of profits or royalties. Licensing an invention is not tantamount to assigning or selling patent rights. A license does not transfer the patent rights from the patent holder to another party. Licensing is like renting rather than selling (Derwent 2001b). Exclusive licenses grant the licensee exclusive rights to the invention; nonexclusive licenses grant licensees nonexclusive rights. In a nonexclusive arrangement, the inventor may grant several different parties rights to make, use, or commercialize his invention. Nonexclusive licenses tend to be less profitable for the licensee because they do not allow the licensee to prevent others from making, selling, using, or importing the invention (Foster and Shook 1993). One reason why no companies decided to take advantage of the NIH's controversial DNA patents discussed in chapter 1 is that the NIH was not granting exclusive licenses. Cross-license agreements (CLAs) are contracts between two or more patent holders that allow the parties to use, make, or commercialize each others products, in exchange for royalties. Reach through license agreements (RTLAs) are contracts between two or more patent holders allowing the parties to use each other's inventions to develop new (or downstream) inventions or products in exchange for royalties. In most industries, companies sign licensing agreements in order to avoid costly and troublesome patent infringement litigation (Foster and Shook 1993).

It is important to note that U.S. patent laws do not require patent holders to develop, sell, use, or license their inventions. Although companies often try to obtain patents in order to use, make, sell, or license products, it may sometimes be more profitable for a company to not exercise any of these legal rights and to not develop the invention or allow anyone else to develop it. The invention would be blocked until the patent expires. Indeed, many companies apply for patents for the sole reason of keeping the invention (or downstream technologies dependent on the invention) off the market. Other countries have a compulsory licensing system in which patent holders must make, use, or market their inventions or license others to do so.

However, the U.S. government does have the power to practice its own form of compulsory licensing if it needs a patented invention for some important state interest, such as satellite communications, electronic encryption, or military technologies. The federal government may override any patent owned by a private citizen if it provides the owner with reasonable compensation for his loss (Resnik and De Ville 2002). Although the U.S. government has not used this power very often, it has a right of eminent domain when it comes to all property rights, including patent rights. This doctrine of eminent domain is typically used to condemn buildings or acquire land for public projects, such as highways (Barron and Dienes 1999). Although state governments do not have the same privileges vis-à-vis patents as the federal government, states are immune from patent infringement lawsuits under the eleventh amendment to the Constitution, unless they exhibit a pattern of infringement (*Florida Prepaid Postsecondary Ed. Expenses Bd. v College Savings Bank* 1999).

Patent infringement occurs when an individual or organization uses, makes, commercializes, or imports a patented invention without the patent holder's permission. Infringement may occur when the infringer makes a new invention that infringes on an already patented invention. A party may infringe on a patent by acting as a direct, indirect, or contributory infringer. A direct infringer is one who makes, uses, or commercializes the device without permission; an indirect infringer encourages a direct infringer to infringe, and a contributory infringer provides materials whose sole use is to make the infringing device (Miller and Davis 2000). In the United States, the standard for determining whether a new invention infringes on a previous one is functional equivalence: if the new invention performs substantially the same function as the old one, then it infringes on the patent (Miller and Davis 2000).

For example, if a pharmaceutical company develops a drug that is functionally equivalent to, though structurally somewhat different from, a patented drug, the new drug infringes on the patent. A "me too" drug is a drug that performs the same function as an existing drug, such as lowering blood pressure. To patent a "me too" drug, a company would need to show that the drug performs a different function or that it is a significant improvement over the patented drug (e.g., that it has fewer side effects). If the company does not show that its drug performs a different function or is an improvement, then it must wait until the patent expires to produce a generic version of the patenting drug. Generic drugs cost less than patented drugs because generic drugs are no longer protected by a company's exclusive control of the drug (Goldhammer 2001).

Infringement can also occur when someone makes or uses a patented part in developing a commercial product. For example, if one company has a patent on some parts, it can refuse to allow another company to use those parts in a new invention (*Telecomm Technical Services, Inc. v Siemens Rolm Communications, Inc.* 2000). The courts will generally allow a company to prevent other inventors from using its patented parts, unless the court determines that the company is abusing its patent rights and extending them beyond their proper scope (*United States v White Motor Co.* 1961; *Eastman Kodak Co. v Image Technical Services, Inc.* 1992). For example, a patent on a transistor should give the inventor exclusive rights over that transistor but not exclusive rights over the entire electronics industry. Although patents give patent holders a monopoly on a particular product, one can still apply antitrust laws, which are designed to promote fair competition patenting practices, to patents (Lao 1999). Patents on parts are especially important in the electronics industry as well as in the biotech industry, as we shall see in chapter 7, because patents on parts (and basic tools) can have downstream effects on research and development.

In the United States, patent holders are responsible for notifying the PTO of possible patent infringement. In infringement cases, patent holders usually inform the infringing party that they are infringing on a patent and that they should stop immediately. If the infringement continues, the parties may settle out of court and may reach a license agreement. Most companies wish to avoid patent infringement cases to avoid costly and time-consuming legal quagmires. If a case goes to court, the court will order a temporary injunction that forbids the accused party from continuing their disputed conduct until the case is settled. In the United States, patent infringement cases are heard in federal courts. If the court rules in favor of the patent holder, the infringing party may have to pay a hefty fine as well as damages. In one of the more famous DNA patent infringement cases, the University of California (UC) spent over $20 million in legal fees in a nine-year effort to prove that

Genentech infringed on its patent on the gene for human growth hormone by making a synthetic version of the compound, Protropin. After many different hearings and appeals, UC agreed to settle the case for $200 million. UC had sued for $4 billion in damages (Barinaga 1999).

One of the important issues related to potential patent infringement concerns the scope of the patent. A patent application contains a description of the invention as well as claims about the invention. When the PTO grants a patent, it grants the inventor exclusive rights with respect to what is claimed in the patent application, including revisions to the document during the application process (Miller and Davis 2000). In an infringement lawsuit, a court must interpret the scope of the claims made in the patent application. In interpreting the scope of the patent, the court confronts the issue of functional equivalency, since the scope of the patent will cover all those inventions deemed to be functionally equivalent to the invention. Frequently, the patent claims made in an application are ambiguous and the courts must interpret the scope of the patent. In making these determinations, the courts therefore confront important policy questions, since construing patent claims broadly increases the control inventors have over a potential field of research and development and therefore also increases their economic incentives, while construing claims more narrowly decreases control as well as economic incentives (Vacchiano 1999; Ko 1992). In general, the courts will give an invention that stakes out a new field, a pioneer invention, a much broader scope than an invention that comes along after the pioneering invention (Miller and Davis 2000). The rationale for this policy is that it rewards innovation instead of trivial modifications.

Courts in the United States (and in Europe) have developed an important, though rarely used, exemption to patent infringement known as the experimental use (or research) exemption. Although the exemption has no basis in statutes, some cases have recognized the exemption as a defense to an infringement lawsuit. The U.S. research exemption dates to early 1800s, when the Federal Circuit Court of Massachusetts held that a defendant had not infringed a patent because the alleged infringing use was for research and not for profit (*Whittemore v Cutter* 1813). Since the ruling in *Whittemore,* the state courts have narrowly construed the research exemption and have not clarified what it means for a use to be for research and not for profit (Karp 1991). In *Roch Products v Bolar Pharmaceutical Co.* (1984), the federal courts considered the experimental use defense for the first time and ruled that the defendant's use did not qualify as experimental. The defendant was a pharmaceutical company planning to market a generic version of a patented drug. The company began some limited testing of its generic drug before the patented drug's patent had expired. The court held that this use was not experimental because it was for business reasons and not strictly for research or scientific inquiry (*Roch Products v Bolar Pharmaceutical Co.* 1984). As one can see, a key point of contention in applying this doctrine is defining research or experimental use and business use.

The research exemption makes a great deal of sense when one is able to distinguish clearly between commercial and noncommerical uses of inventions. The general idea behind the exemption is that academic scientists should be able to use patented inventions to conduct basic research without fear of infringement, but that industrial scientists conducting applied research or product development should obtain permission from the patent holder to avoid infringement. The research exemption could fill an important niche in patent law similar to the fair use exemption in copyright law (discussion follows) (Gitter 2001). Like the fair use exemption, a key test would be whether the use of the invention diminishes the economic value of the invention.

Unfortunately, the research exemption doctrine, as it is currently understood, is vague. To aid the courts in interpreting patent law, it is important to have a clear statement of the doctrine that defines its keys terms. In many fields of research today, it is difficult to distinguish between commercial and noncommercial research. First, noncommercial basic research conducted in universities often has immediate, economically significant applications. Indeed, universities now vigorously pursue their intellectual property rights and encourage researchers to apply for patents. A typical university will provide faculty members with legal assistance in obtaining patents in return for a fifty-fifty split of royalties from licensing. Universities also usually assert that they have intellectual property interests in any invention conceived by faculty employees (Resnik and Shamoo 2002). Second, many private companies conduct basic research inhouse or fund basic research offsite in academic settings. For example, we saw in chapter 2 that Celera Genomics has funded a great deal of basic researchers on DNA sequence data for the human genome. Celera has also sponsored research on the fruit fly and mouse genomes (Marshall 1999b). Third, many universities and private corporations have developed partnerships, including research institutes. Thus, the line between commercial and noncommercial research is often very blurry in today's research environment. Because this line has blurred, even the most precise definitions and rules may be difficult to apply. To overcome these ambiguities, some have suggested that governments codify the research exemption by enacting statutes that define the exemption (Nuffield Council on Bioethics 2002).

Finally, it is worth mentioning the under U.S. (and European) law a third party can challenge a patent after it has been issued by the patent agency. In the United States, a third party can request a reexamination of the patent, which can take place at any time during the life of the patent. To succeed in challenging a patent, one must produce evidence that the invention had already been disclosed in the prior art. Challenges to patents are rare, and those that lead to reexamination rarely succeed. More often than not, challenges result in an amendment to the patent rather than revocation of the patent (Nuffield Council on Bioethics 2002).

COPYRIGHTS

A copyright is legal protection that gives the holder the right to control the reproduction, modification, performance, and public display of an original work such as software, essays, books, plays, paintings, movies, songs, databases, or sculptures. Just about any form of expression may be copyrightable. The key test is whether the expression has been fixed in a tangible form. Thus, a person's oral speech is not copyrightable because it is not tangible. A transcript of the speech is copyrightable because the transcript in tangible. The tangible expression does not need to be something that a human being can directly perceive: under U.S. law, phonograph records and cassette tapes are copyrightable, even though a person can only indirectly perceive these recordings (Miller and Davis 2000).

Copyright holders also have rights pertaining to the reproduction of works derived from their original work. Normally, the author(s) of the original work holds the copyright on the work, unless the work is a work for hire, in which case the person (or agency) who contracts for the work owns the copyright. A person who has copyrights over an original work may transfer those copyrights to someone else or license others to copy the work. For example, most software sold today contains a licensing agreement in which the purchaser agrees to certain terms of use of the software (i.e., the purchaser may not make unauthorized copies of the software).

Copyrights are renewable and last much longer than patents: in the United States the length of copyright protection is the lifetime of the author plus seventy years. The protection for works for hire is ninety-five years from publication or one hundred and twenty years after creation (Miller and Davis 2000; Foster and Shook 1993). If a copyright is not renewed or expires, the work becomes part of the public domain and can be copied without permission. Although U.S. laws permit and encourage copyright registration through the U.S. Copyright Office, all authors of original works have full copyright protection even if they do not register their work. Registration is only necessary to pursue a copyright infringement lawsuit (Davis and Miller 2000).

One key requirement for acquiring a copyright is originality: to obtain copyright protection over an original work, the author must have been the first person to create the work. The author must not have copied the work from someone else; the work must be his or her own work and not someone else's. Copyright law, unlike patent law, does not require that authors show that their work is novel or nonobvious. Indeed, one may copyright a work that is not novel and is completely obvious, such as a reproduction of an original work. To obtain a copyright on a reproduction of someone else's original work, one must add something to the work. Thus, one may not copyright a reproduction of a painting if the reproduction introduces only nontrivial variations (*Alfred Bell & Co. v Catalda Fine Arts, Inc.* 1951). One may even copyright a compilation of facts (or a database), provided that one has made some minimal creative contribution to the compilation of facts, such as a creative way of presenting or organizing the facts (*Feist Publications v Rural Telephone Service* 1991). The key requirement for copyright protection is that the author has added something to the work that makes the work his work as opposed to someone else's work or a collection of facts.

Another essential requirement for a copyright is lack of utility: one may not copyright a useful and patentable invention (Miller and Davis 2000). For instance, if a lamp contains parts that are useful as well as parts that are merely ornamental (and serve no practical use), one may copyright only the nonuseful parts of the lamp (*Mazer v Stein* 1954). The rationale for not allowing people to copyright useful inventions is that patent laws protect items that have utility and patents expire in twenty years. If one could copyright a useful invention, this would keep the invention out of the public domain for seventy or more years.

Finally, to obtain a copyright there must be no merger between the idea and the expression. If the tangible expression is virtually the only way of representing an idea, then the expression merges with the idea and one is barred from obtaining a copyright. The rationale for this doctrine is that copyrights protect forms of expression but not ideas (Miller and Davis 2000). For example, the U.S. Supreme Court held that a one may copyright a document describing a system of accounting, but that one may not copyright the system of accounting itself. A person could use that system of accounting without infringing copyrights, although a person could not copy the document describing the system without infringing copyrights (*Baker v Selden* 1879). Thus, a person does not have the right to control the facts or ideas expressed in the original work. Someone who takes a photo of the moon can copyright their photo but not the moon itself. Likewise, the author of a scientific paper can copyright the paper but not the data reported in the paper. Someone who compiles a database can copyright the database but not the facts in the database.

Copyright infringement occurs when someone uses some of the exclusive rights over any original work, such as the right to copy, without permission of the copyright holder. To pursue an infringement action, the plaintiff can bring his case to a federal court in which he must prove that he is the owner of the copyright and that impermissible use (e.g., copying) occurred. It is rarely the case that the plaintiff will have direct evidence the defendant copied the work, so he must usually appeal to circumstantial evidence, such as a demonstration of an extraordinary resemblance between his work and the unauthorized copy

(Miller and Davis 2000). As an added deterrence to copyright violation, the United States has enacted criminal penalties for willful copyright infringement for commercial purposes. According to the Federal Sentencing Guidelines, criminal copyright infringement is punishable by fine or one to ten years in prison (Miller and Davis 2000).

Copyrights also allow for some unauthorized use of parts of the original work under the doctrine of fair use. Fair use is usually interpreted as a defense to copyright infringement: the infringer must demonstrate that his use of the copyrighted work was fair (Dreyfuss and Kwal 1996). To determine whether a use is fair, the court examines the following (Miller and Davis 2000):

1. The purpose of the use. Was the use for a commercial or noncommercial purpose?
2. The nature of the copyrighted work. Was the copyrighted work a commercial or noncommercial (e.g., academic) work?
3. The proportion of the work used. How much of the work was used?
4. The economic impact of the use. Did the use negatively impact the economic or commercial value of the work?

Although the courts consider all of these factors in deciding whether an unauthorized use is a fair use, the last factor, economic impact, is by far the most important. The rationale for the doctrine of fair use is twofold. First, it allows for some unauthorized uses of the work and thus gives the public greater use of the work. Second, when authors make their works available to the public they assume that people will use their works; authors implicitly consent to some unauthorized uses. The courts have found a variety of unauthorized uses to be fair uses, including home videotaping of a television program for personal use, limited photocopying of academic works for educational or research purposes, and parody, commentary, satire, and news reporting.

Since the courts weigh several different factors in determining fair use, it is often hard to predict whether the courts will find that a particular use is a fair use. The doctrine is so difficult to interpret and apply that many attorneys find it to be extremely vexing. One court described it as "the most troublesome in the whole law of copyright" (*Dellar v Samuel Goldwyn, Inc.* 1939, 662). One reason why the doctrine is so difficult to interpret and apply is that it requires courts to balance the interests of the author in protecting his original work and the interests of society in using it (Dreyfuss and Kwal 1996). As noted earlier, this dilemma echoes throughout all of IP law. Although the patent system has no doctrine of fair use, it has a research exemption, which courts have used infrequently. One wonders whether a research exemption doctrine, if it became an influential part of patent law, would become like the doctrine of fair use. In order to avoid some of the problems that have plagued the doc-trine of fair use, countries would be wise to enact statutes that clearly define the research exemption, if they plan to use this in patent law (Nuffield Coun-cil on Bioethics 2002).

One relatively recent development in copyright law concerns copyright protection for electronic databases (Gardner and Rosenbaum 1998). Online, electronic databases play a crucial role in nearly all areas of research and schol-arship, ranging from genomics and medicine to marketing and the law. As we noted in the first chapter, Celera Genomics has developed an electronic genomics database. Researchers, companies, and government agencies have developed many different databases pertaining to genetics, genomics, pro-teomics, and structural biology (Roberts et al. 2001). Researchers and schol-ars, as well as private businesses and government agencies, benefit greatly from having quick and easy access to huge quantities of information in a usable and searchable form. Students, teachers, patients, and consumers also benefit from access to these databases.

Since it is very easy to copy, rearrange, and edit these databases, elec-tronic databases raise difficult issues about intellectual property protection. Many private companies, such as the National Association of Realtors, Lexus/Nexis, Celera Genomics, and eBay, have invested billions of dollars in compiling electronic databases and they do not want their works to be pirated. Electronic databases are compilations of information from various sources. In a key case relating to databases, *Feist Publications v Rural Tele-phone Service Co.* (1991), the U.S. Supreme Court said that a telephone book could not be copyrighted because it was not an original compilation of facts. The Court also ruled that a compilation of facts can be copyrighted if it has some minimum degree of originality. It is likely that many, though not all, of the electronic databases used in scientific research would have copyright pro-tection under the ruling in *Feist*, since these databases probably have some original arrangement or presentation of facts. Even if a database is protected by copyright, the doctrine of fair use would still allow individuals to copy parts of the database without permission, if that unauthorized copying does not affect the commercial value of the database. Since copyright laws do not protect facts, one may always use the facts compiled in the databases without obtaining permission from the database compiler. Since copyright laws still allow for a great deal of unauthorized uses of databases, databases owners have lobbied for new laws to provide strong database protection, while researchers, librarians, consumers, and other groups with interests in access to information have lobbied against strong database protection and in favor of weaker database protection (Gardner and Rosenbaum 1998). The issue of database protection raises some perennial questions in intellectual property law and policy: what are the best ways to balance private ownership and rights against public access and use?

TRADEMARKS

Trademarks are the third major type of intellectual property protection recognized by governments around the world. A trademark is a name, phrase, image, or symbol a business uses to distinguishes its goods or products (e.g., the name Microsoft, McDonald's golden arches, the Planters peanut man, or the phrase, "it's the real thing"). The defining feature of a trademark is distinctiveness: a trademark will obtain protection if it is distinctive (i.e., if it allows consumers to distinguish the goods of the trademark applicant [or holder] from someone else's goods). The owner of the trademark has the right to exclude others from using the trademark without permission for an unlimited time, that is, as long as he is using his trademark (Foster and Shook 1993). Since the Constitution does not address trademarks, trademarks, unlike copyrights and patents, do not enjoy strong protection in federal law. In the United States, state laws protect trademarks and the federal government has a trademark registration system. Although trademark registration does not confer federal rights on the trademark holder, it does serve as constructive notice to other businesses that the trademark holder is using the trademark, which will be useful in overcoming the defense of innocent infringement in a lawsuit (Miller and Davis 2000). To prove that someone has infringed a trademark, the plaintiff must show that consumers are likely to be confused by the unauthorized use of a mark similar to the one used by the plaintiff. Trademarks are very important in business and industry but not very significant in scientific research. It is not likely that trademark disputes will arise in genomics research, unless a company decides to use a gene as a trademark or trademarks the name of a gene. Therefore, this book will not discuss this topic any further.

DNA AS INTELLECTUAL PROPERTY

Now that we have set forth some basic principles of intellectual property law and policy, we can begin to tackle the complex and controversial issue of DNA as intellectual property. If you are a scientist, inventor, or biotechnology company interested in having proprietary control over DNA, what would be your best strategy? What strategy would the law permit? What strategy would make the most economic sense? In some cases, trade secrecy might be a viable option. If you have developed a process or method in biotechnology that has economic value and you can keep it a secret, then you might protect your interest through trade secrecy. But this strategy is not a very effective way of protecting genetic information for the simple reason that it is relatively easy for a competitor to make an independent discovery of a DNA sequence or use standard techniques in biotechnology to reverse engineer the sequence. If you isolate and purify a DNA sequence from a naturally occurring species, another researcher with the necessary skills and knowledge can perform the same task. Thus, trade secrecy is often not very useful. Trade secrecy may be more important in the future, when it is possible to design new DNA sequences, proteins, and structures from natural products and processes. But even then, these new inventions may succumb to reverse engineering (Resnik 2001a).

Copyrights also present problems for those seeking property rights in DNA. Recall that copyright law applies to original works, not to the facts or ideas represented by those works. Copyright law also does not apply to items that have a utility. Both of these issues would present legal problems with copyrighting a DNA sequence. First, since DNA sequences are chemical compounds that exist in nature, attempting to copyright a sequence would be like trying to copyright a fact, such as any other chemical formula. A person could copyright a paper describing a chemical formula but she cannot copyright the formula itself; copyright law protects tangible expressions, not facts. Second, since DNA sequences are useful, attempting to copyright a DNA sequence would be like attempting to copyright any other useful chemical compound, such as Teflon or Styrofoam. Since these compounds are useful, then are not copyrightable. Likewise, DNA is useful (a point we will discuss later). It is patentable but not copyrightable.

Thus, while one can copyright a paper or computer program describing a DNA sequence, one cannot copyright the sequence itself.[3] Since the sequence cannot be copyrighted, anyone else can make a copy of the sequence without violating copyright law, just as anyone can take a photograph of the moon without violating someone else's copyright of a photograph of the moon. On the other hand, as we saw earlier, copyright law affords some protection for databases that have some minimal standard of originality. Thus, it should be possible for companies, such as Celera Genomics, to obtain copyright protection of genetic information compiled in databases. Celera has published its own versions of the human and mouse genomes and planned, at one time, to make most of its income from selling genomics information services (Celera 2001). However, it is likely the copyright protection, though important and useful in biotechnology, will not be nearly as lucrative as patent protection (Resnik 2001a).

DNA PATENTING

The alternative to trade secrecy and copyright is patenting. Most of the IP activity in DNA and biotechnology has focused on this type of protection. It has been legal to patent biological materials in the United States since 1980. Prior to this time, the United States did not recognize patents on biological materials, including plants and animals, because the courts treated animals and plants as products of nature, not as products of human ingenuity, although hybrid plant varieties were a notable exception to this policy. The United States enacted several laws permitting the patenting of hybrid plants, including the Plant Patent Act, the Plant Protection Act, and the Plant Variety Protection Act (1994) (Kevles 2001). The courts affirmed this policy of forbidding biological patents in *Funk Brothers Seed Co. v Kalo Inoculant Co.* (1948), in which the Supreme Court ruled that naturally occurring plants are not patentable, although hybrids are patentable.

The United States granted the first DNA process patent to Stanley Cohen of Stanford University and Herbert Boyer of the University of California, Berkeley for a laboratory method for cloning recombinant DNA (Cohen and Boyer 1980). The Cohen-Boyer patent played an important role in the development of recombinant DNA technology and helped to launch the biotechnology industry. By the fall of 1977, Genentech, then a new company, had used the methods described in the patent to create bacteria that manufacture human proteins. Although many scientists had suspicions about the commercialization of biotechnological research, it soon become apparent to researchers, including Boyer, that recombinant DNA had reached the point where it would have commercial applications (Hughes 2001). Boyer, who helped to found Genentech, was featured on the cover of *Time Magazine* on March 9, 1981. The caption read: "Shaping Life in the Lab: The Boom in Genetic Engineering."

Although the Cohen-Boyer patent played an important role in providing the legal foundation for DNA process patents, it did not establish the patentability of DNA products. The patent did not really cover any radically new legal ground, because the patent office had awarded patents on the use biological process prior to the Cohen-Boyer patent, such as patents related to fermentation (see Eweson 1976; Nickel 1976). However, a patent filed by molecular biologist Ananda Chakrabarty, which did cover new legal ground, eventually wound up in the Supreme Court.

In the early 1970s, Chakrabarty used recombinant DNA technology to transfer a gene to a bacterium that allowed the organism to metabolize crude oil. Chakrabarty did not think that he had invented anything, but he sought legal advice and filed a patent on June 7, 1972. The patent examiner at the PTO turned down his claim, citing existing legal rulings that naturally occurring species are not patentable. Chakrabarty appealed his case to the Patent Office Board of Appeals, which affirmed the patent examiner's decision. Chakrabarty appealed his case to the Court of Customs and Patent Appeals, which held that his invention was patentable. The Commissioner of Patents, John Diamond, then appealed the case to the Supreme Court, which ruled in favor of Chakrabarty by a 5 to 4 decision and affirmed the ruling of the Court of Customs and Patent Appeals. Chief Justice Warren Burger delivered the opinion of the Court. The American Society for Microbiology, the Pharmaceutical Manufacturers Association, and the Regents of the University of California filed amicus briefs in the case.

The key legal issue before the court was whether Chakrabarty's bacteria constituted a manufacture or composition of matter under the U.S. Patent Act (1995). In reviewing the legislative history of the Patent Act, the Court noted that the author of the Act, Thomas Jefferson, supported a broad construal of the subject matter of patents. Jefferson, according to the Court, believed that ingenuity should receive liberal encouragement. It also mentioned that a Congressional Report on the 1952 revision of the Patent Act said that "anything under the sun that is made by man" should be patentable (*Diamond v Chakrabarty* 1980, 307). The Court also indicated, however, that not every discovery should be patentable; laws of nature, natural phenomena, and abstract ideas should not be patentable. The Court applied this reasoning to the case and determined that Chakrabarty's bacteria should be patentable because it is a product of human ingenuity that differs markedly from naturally occurring bacteria. The bacteria is not "nature's handiwork, but his own" (*Diamond v Chakrabarty* 1980, 308).

In arriving at this holing, the Court also rejected an argument made by the petitioner (Diamond) that microorganisms cannot be patented because Congress did not authorize this application of the Patent Act, since it could not foresee the technological developments that had taken place since the last revision of the Act. According to this argument, the Court would be overstepping its bounds by allowing an organism to be patented, and only Congress can make this decision. The Court acknowledged that Congress has the authority to define the limits of patentability, but it also stated that it must interpret statutes in terms of their legislative purpose and history. The legislative purpose and history of the Act, according to the Court, supports broad construal of key terms like composition of matter and article of manufacture (*Diamond v Chakrabarty* 1980, 310).

The Court also rejected another argument made by Diamond and several people who submitted amicus briefs that allowing organisms to be patented would unleash a "gruesome parade of horribles," including environmental pollution, the spread of disease, and the deprecation of the value of human life (*Diamond v Chakrabarty* 1980, 311). Though the Court recognized the risks inherent in biotechnology, it also said that granting or denying patents would probably not stop genetic research or protect society from its risks. The Court also noted that these broad concerns about the risks of biotechnology are matters of high policy, which are best addressed through legislative processes.[4]

Shortly after the Court reached its decision, the PTO awarded Chakrabarty his patent for "Microorganisms Having Multiple Compatible Degradative Energy-Generating Plasmids and Preparation Thereof" (Chakrabarty 1981). Since the Court's ruling in *Chakrabarty*, the PTO has awarded thousands of patents on biological products, including patents on genes, SNPs, ESTs, cell lines, mice, plants, rhesus monkeys, and human stem cells. Although the PTO had allowed patents on biological processes prior to the Chakrabarty case, the Supreme Court's decision established a legal precedent for patents on biological products (Eisenberg 1990, 1995). From 1990 to 1999, the number of DNA patent applications received by the PTO rose from less than 500 to almost 2700 per year (Enserink 2000). As of November 2001, the PTO had awarded 50,099 patents related to DNA, 58,362 patents related to genes, 95,898 patents related to proteins, 283,301 patents related to cells, and 70,425 patents related to mice.[5]

While the Chakrabarty decision can be construed as supporting a patent on a novel DNA sequence, one might wonder why it also supports patents on cloned DNA sequences, which have natural counterparts. A cloned DNA sequence is like a copy of a natural sequence. The PTO has held that cloned DNA sequences are patentable if these sequences are regarded as isolated and purified chemical compounds (Doll 1998, 2000). One of the key ideas in patent law is the doctrine of isolation and purification. Many chemicals that occur in a natural state, such as digitalis in the foxglove plant, can be isolated from nature and purified (or processed) to produce a compound, such as digoxin, that does not occur in nature. The isolated and purified product is a human invention, although the product in its natural state is not (Doll 1998, 2000). Likewise, DNA in its natural state is not a human invention, but isolated and purified DNA is a human invention. (We will examine this point in more depth in chapter 5.)

There are three basic types of patents that the PTO has issued related to DNA: product patents, process patents, and improvement patents (Eisenberg 1997). A product patent is a patent on DNA treated as a composition of matter or article of manufacture, such as a particular isolated and purified or modified DNA sequence. A process patent is a patent on processes or methods related to DNA, such as process for isolating, analyzing, sequencing, or making DNA. An improvement patent is a patent on an improved DNA sequence, such as a modified sequence, or an improved DNA process.

For example, Amgen owns a patent on an isolated and purified DNA sequence that encodes the erythropoietin (EPO), a protein that stimulates the production of red bloods cells (Sagoff 1998). The patent gives Amgen exclusive rights to use the gene to manufacture EPO or to license others to do so. Amgen also has patents pertaining to processes for manufacturing EPO and processes for isolating and purifying the EPO gene. To get the fullest possible patent protection, Amgen has obtained product and process patents related to EPO and the EPO gene. Myriad Genetics' controversial patents on breast cancer genes assert claims to methods for detecting genes related to breast cancer, but are not composition of matter patents (see, for example, Shattuck-Eidens et al. 1997). To increase its intellectual property protection with respect to genetic tests, Amgen also obtained patents on methods for detecting mutated proteins associated with breast cancer (Adams et al. 1999). Although DNA can be patented so long as it falls into one of the basic categories of patents, many companies seek to increase their patent protection by filing for many different types of patents.

Most of the controversy surrounding DNA patents has focused on product patents, since these patents give inventors exclusive rights relating to isolated and purified DNA sequences. As noted in Chapter 1, critics of DNA patents have argued that product patents are, in effect, patents on nature. We will explore this objection in more detail in Chapter 5. Others have objected to DNA product patents on the grounds that they do not meet the legal conditions for patenting, such as novelty, usefulness, and nonobviousness. We will discuss these issues shortly. Process and improvement patents, on the other hand, have generated far less controversy, because most people recognize that the processes related to DNA and improvements on DNA are human inventions, which meet the other requirements for patenting, such as novelty, nonobviousness, and utility. This book will therefore focus most of its discussion on DNA product patents.

LEGAL ISSUES RELATED TO DNA PATENTS

If we apply our earlier general discussion of patenting to the case of DNA patenting, we can see that the courts must address the following questions to determine if DNA patents are valid:

1. Is DNA a product, process, or improvement on a human invention? Is it a patentable subject matter?
2. Are DNA inventions novel? Are they barred by prior disclosures or uses?
3. Are any DNA inventions invalid as a result of statutory bars?
4. Are DNA inventions nonobvious? Would a person trained in the prior art be expected to be able to make a DNA invention?
5. Are DNA inventions useful? What practical uses do they have?
6. What must be disclosed in a DNA patent application?
7. What is the scope of DNA patent? What should inventors be allowed to claim in a DNA patent?

Although these questions raise legal issues, which can be answered through careful, legal scholarship and analysis, they also raise moral, social, political, and theological issues, which cannot be answered through legal inquiry alone. The answers to these questions about DNA patenting presume particular views about DNA, living things, human invention, and ownership. For example, the PTO views DNA as a chemical compound, but some people view DNA as more than a mere chemical; they view it is a chemical with cultural, moral, and theological significance. The PTO regards biological organisms as complex compositions of matter, which are reducible to patentable parts. Others view organisms, including human beings, as unified wholes that are not reducible to patentable parts. We will explore these questions in chapters 5 and 6.

The bulk of this book will focus on inquires related to question 1, since most of the controversy in DNA patenting has revolved around this question. Also, this question is the most fundamental one, since if DNA is not a patentable subject matter, then all the other questions are moot. I will return to question 1 later in the book, but I will consider the other questions presently.

Questions 2 and 3 are, for the most part, technical, legal questions that do not raise many deeper issues. Thus, this book will not spend much time on these questions.

Question 4 looks like a technical, legal question, but it raises some important social and policy issues related to the determination of obviousness. When people isolated and purified the first DNA sequences, it required a great deal of technical skill and effort to achieve this result. Since a DNA sequence would not be obvious to a person trained in the relevant art, it would meet the nonobviousness requirement. However, advances in genomic analysis, biotechnology, and bioinformatics have routinized DNA sequencing. Automated sequencing machines can produce thousands of DNA sequences per day, and scientists and technicians can use powerful computers and databases to determine their structure and protein products. Thus, one could make the case that DNA sequences are now obvious, even if they were not obvious years ago and that patent offices and the courts should raise the bar for obviousness (Dastgheib-Vinarov 2000; Nuffield Council on Bioethics 2002). *In re Deuel* (1995) addressed some of these issues. In this case, the court ruled that a patent on a DNA sequence was not rendered obvious by the mere fact that it was created using standard techniques. What matters in determining obviousness is the state of the prior art, not advances in methods for making DNA.

Even if one accepts the argument, rejected in *In re Deuel*, that advances in DNA biotechnology make DNA sequences obvious, many DNA patents would still be able to meet this higher nonobviousness hurdle. Although the DNA sequence may, itself, be obvious, an unexpected modification of a DNA sequence will not be obvious. Thus, this critique would undercut the legality of patents on isolated and purified sequences but it might have little effect on patents on modified sequences. Second, even if the sequence is obvious, uses of the sequence may not be obvious. For example, a patent on a DNA sequence used in predicting breast cancer risks would not be obvious. Even if the sequence is obvious, its use in predictive genetic testing would not be obvious. For comparison, anyone can take regular newspaper, tear it to pieces, mix with it sawdust (or wood shavings) and used motor oil, and compress it into a brick. This much is obvious. What is not obvious is that one might use this mixture as a quick-starting and long-burning log for campfire. It is important for patent offices and the courts to raise question 4, since it will require inventors to come up with inventions that are not obvious. Clearly, society does not benefit when the government awards patents on obvious inventions. Those who wish to patent DNA should take appropriate steps to show that their inventions are not obvious.

Moreover, what counts as obvious is not obvious. To determine whether an invention is obvious, one must examine the prior art and compare the invention to the prior art. In making this examination, one must make many different interpretive judgments that affect the balance between private ownership and public access: if one sets a low bar for the obviousness standard, then this favors private interests, since it is more likely that inventions will meet this standard. Conversely, if one sets a high bar, this favors public access, since it is less likely that inventions will meet this standard. Since setting the standard for obviousness in patent law has important implications for public policy, patents offices and the courts will need to find a way to achieve the best balance of private control versus public access.

As mentioned in chapter 1, question 5 has generated some controversy in recent years as well and the PTO recently took steps to raise the bar for proving utility. The PTO took these steps in response to abuses of the patent system in which researchers filed patents on DNA sequences without stating a definite, precise, and nonspeculative use for those sequences. I have no qualms with PTO's decision, but I would like to ask a deeper question about the usefulness standard: what does it mean for a DNA sequence to be useful? Patent law does not provide us with a general definition of usefulness beyond the idea that to be useful something must be (a) not frivolous and (b) not harmful to the public (*Application of Nelson* 1960, 175). Courts tend to take a very liberal interpretation of usefulness and focus instead on evidentiary requirements placed on patent applicants for proving some specific use.

To get a better understanding of what is means to be "useful," we would do better to conduct a philosophical analysis of how we typically use the words useful and use. Let's start with some dictionary definitions. The *American Heritage Dictionary of the English Language* (2000) defines the verb "to use" as "1. To put into service or apply for a purpose; employ" and "4. To seek or achieve an end by means of." It defines "useful" as "1. Having a beneficial use," and "2. Being of practical use." Thus, to be useful, a thing must be capable of meeting or serving some purpose or goal. For instance, a toothbrush is useful for achieving the goal of brushing teeth. Brushing teeth, in turn, is useful for the goal of having clean teeth, which is useful for the goal of achieving dental health. The general point is the usefulness is a teleological (or goal-directed) notion: an object, A, is useful with respect to some goal, G.

Function is a synonym for the word use in the sense that both words can designate a role played by some object in a system. Many people use a toothbrush to brush their teeth. Thus, a function of a toothbrush is to brush teeth. A brick can function as a doorstop; thus, a brick can be used as a doorstop. The concept of function, like the concept of use, is a teleological notion: an object, A, has a function with respect to some goal, G, in a particular system, S (Nagel 1977). There are different types of functions, including biological functions and human (or artificial) functions (Rosenberg 1985). For example, the function of the heart is to pump blood because its presence in organisms can be explained by referring to its ability to meet a particular goal (i.e., pumping blood). Natural selection has designed hearts to meet this goal. Organisms with hearts that pumped blood effectively were able to outperform organisms with ineffective hearts in the game of evolution. The function of a traffic light, on the other hand, is to regulate traffic flow because its presence can be explained by referring to human goals or intentions: traffic lights exist because human beings designed them to meet goals related to traffic regulation. The difference between a biological function (or a biological use) and a human function (or human use) boils down to this: biological functions serve biological goals, such as reproduction and survival, while human functions serve human goals, such as traffic regulation, transportation, entertainment, and so forth (Rosenberg 1995). In both cases, the concepts are teleological; they differ, however, with respect to their goals.

Can these distinctions help us to understand the usefulness of DNA? Under a biological approach to function and use, DNA sequences that code for proteins (codons) have functions (or uses): to carry information required for making proteins. They would have these biological functions even if human beings did not exist to discover these functions. Other DNA sequences, such as junk DNA, may not have biological functions and may not be useful (form a biological point of view). Thus, if a DNA sequence codes for a particular protein, one can always say that its function or use is to carry information for making the protein. All one would need to do to assign the sequence a use would be to determine what the corresponding protein happens to be, which is an inquiry that is becoming much easier as result of advancements in genomics, proteomics, and bioinformatics.

Would proving that a DNA sequence codes for a specific protein be sufficient for claiming that it has a practical use? This would show that the protein has a biological use but not that it has human use. It should not, therefore, satisfy the utility requirement in patent law. To demonstrate that a DNA sequence is useful one must show how it serves particular, human goals. If a patent applicant can demonstrate that the protein itself has a particular use (e.g., that it can be used to diagnose or treat diseases), then this would be sufficient to show that the DNA sequence has a practical use. Thus, DNA patenting may become closely linked to protein research and patenting.[6] This conclusion, based on philosophical argument, supports the PTO's recent communications about its interpretation of practical utility.

On the other hand, the mere fact that a DNA sequence has no apparent biological function or use does not prove that it has no human function or use. As noted earlier, junk DNA has no apparent biological function, but it may still be useful in DNA fingerprinting and in analyzing genetic variation. Thus, one might be able to show the certain junk DNA sequences meet the utility requirement even if they have no biological uses or functions.

This philosophical digression into questions about function and use also provides us with some insight into question 7, the question about scope. As noted earlier, the key test for determining patent infringement is functional equivalency: an invention infringes a patent only if it performs substantially the same function as the patented invention. More specifically, it must perform substantially the same function in substantially the same way to produce substantially the same result (Miller and Davis 2000; *Warner-Jenkinson Co. v Hilton Davis Chemical Co.* 1997). Substantially is one the great wiggle-words in the law because it can be interpreted in many different ways: what is "substantial" to one attorney or court may be "insubstantial" to another.[7] Since the word substantial can be interpreted in different ways, the courts have considerable ability to affect patent law and policy. For example, a very liberal interpretation of substantial would endow a patent with a very broad scope. Does an incandescent light bulb perform substantially the same function as an ultraviolet one? On a liberal interpretation of substantial, the answer is "yes—both inventions produce light." Thus, an ultraviolet light bulb would violate a patent on an incandescent light bulb. A more restrictive interpretation, on the other hand, would hold that the two inventions do not perform substantially the same function and that the ultraviolet light does not infringe the patent. The general point here is that the word substantial gives the courts some wiggle-room for interpreting the scope of patents.

If we set aside the word substantial for the moment, we see that the word "function" is also subject to a great deal of interpretation as well. What is the function of an incandescent light bulb? One could say that its function is to produce light. Or perhaps its function is to produce light of specified wavelengths. Or maybe its function is to produce heat. Since function is a teleological notion, to specify the function of the light bulb, one must specify the goal(s) that it serves. Two light bulbs will perform the same function if they serve the same goals. But how specific should one be in specifying goals? A broad approach to specifying goals would imply a broad interpretation of scope, since two different inventions could serve the same general goals (i.e., a match and a light bulb could both produce light and heat). On the other hand, a narrow approach to specifying goals would imply a narrow interpretation of scope.

The upshot of this discussion is that the words substantial and function, in patent infringement lawsuits, can be interpreted in different ways. These different interpretations can affect how the courts interpret the scope of a patent. Courts can give patents a broad scope, a narrow scope, as well as a range of values in between these two extremes. When they engage in these interpretations, the courts should consider how their rulings affect the balance of private control versus public access in patent law (Ko 1992; Vacchiano 1999). Courts should attempt to adjust the scope of patents, including DNA patents, in order to strike the best balance between private and public interests (see, for example, *Genentech Inc. v Chiron Corp.* 1997).

I will return to some of issues addressed by questions 1 through 7 later on in the book when I consider their consequences for science, medicine, industry, agriculture, and society.

INTERNATIONAL INTELLECTUAL PROPERTY TREATIES AND LAWS

Before concluding this chapter, it will be useful to address some international issues relating to IP laws. Patents, copyrights, and trademarks are legal protections that were initially only enforceable in the countries in which they were issued. Thus, prior to some international IP agreements, a copyright obtained in the United States would only give the holder legal rights in the United States. This is still the case for patents: a patent granted in the United States only applies in the United States and its territories. However, some international treaties have made copyrights international in scope. Under the Berne Convention, which the United States has not signed, a copyright granted in a signatory country carries legal authority in all other signatory countries and issues to the first person to produce the work in any signatory country. The political difficulty with a treaty like this is that it interferes with national sovereignty in IP laws. Thus, the United States and many countries have resisted international treaties that limit their sovereignty and have signed agreements that commit nations to international cooperation without sacrificing national sovereignty. Most other treaties, such as the Paris Convention, the Uniform Copyright Convention (UCC), and Trade Related Aspects of International Property Rights (TRIPS), do not create international IP rights, but are agreements that address cooperation among signatory countries and international uniformity of IP laws.

For our purposes, TRIPS is the agreement that has the greatest bearing on international aspects of IP. The TRIPS agreement was signed in 1996 by 140 nations who are members of the World Trade Organization (WTO), of which the United States is a member. The TRIPS agreement sets some minimum standards for intellectual property laws as well as principles for international cooperation and enforcement of those laws. If someone holds a patent in two different countries that both recognize the TRIPS agreement, then those countries have agreed not to undermine each other's patent protections through unilateral action. For example, if a pharmaceutical company has a patent in the United States and in South Africa, then South Africa has an obligation to not import generic versions of the drug from countries that do not recognize TRIPS (Resnik 2001b). South Africa would also have an obligation to not use compulsory licensing to produce the drug. On the other hand, the agreement does allow countries to use compulsory licensing to address national emergencies, such as public health crises.

Current international intellectual property treaties do not mention DNA patenting per se, although some international bodies, such as the United Nations Educational, Scientific, and Cultural Organization (UNESCO), the World Health Organization (WHO) and the Human Genome Organization (HUGO), have adopted specific statements about DNA patenting. For example, a 1997 UNESCO declaration condemns profiteering from the human genome in its natural state and regards the human genome as the common heritage of mankind. HUGO and WHO have both declared that those who participate in genetic studies should receive some benefits from participation (HUGO 1995). Although statements by the WHO, HUGO, and UNESCO have some moral and political influence, they lack legal authority. Incidentally, the United States is not a signatory to UNESCO.

Perhaps the most influential international agreements relate to patents in Europe. In 1973, nineteen nations from Europe formed the European Patent Convention (EPC) to strengthen cooperation among these countries with regard to patented inventions. The EPC has been amended several times, most recently in 1998 (European Patent Convention 1998). The EPC has the force of law in Europe and is administered by the European Patent Organization (EPO). Patents granted by the EPO have the same effect among signatories to the EPC as patents issued in those respective countries. The EPO, like the PTO, awards patents based on the applicant's successful demonstration of novelty, nonobviousness, and utility. An invention (defined as something that is susceptible to industrial application) may be patented if it results from an inventive step. The EPC states that scientific discoveries and theories, mathematical methods, and computer programs may not be patented. The EPC also declares an invention is not patentable if its publication or exploitation would be against the public morality (European Patent Convention 1998). This stands in sharp contrast to U.S. law, which lacks a public morality clause (Nuffield Council on Bioethics 2002). Remember, however, that under U.S. law an invention could be found to have no practical use if its only use is a harmful use.

In addition to meeting the requirements of the EPC, European patents must conform to policies set by the European Commission (EC) of the European Union (EU). The EU is an organization that promotes cooperation and trade among its fifteen member nations, all of which are also signatories to the

EPC (European Commission 2001). The EC provides governance and policy direction for the EU. Thus, when the EC issues a policy directive on patents, this directive has clear and direct implications for patent law and policy in Europe. Although any nation is free to ignore the directives of the EC, those that do face possible ouster from the EU. In 1998, the European Parliament and the Council of the EU adopted the "European Directive on the Legal Protection of Biotechnological Inventions," a document that addresses a number of different issues related to the commercialization of genetic research. Although this document does not have the force of national law, it establishes policy standards for EU nations (Knoppers 1999). Since the directive (European Parliament and Council 1998) discusses some key policies relating to DNA and biotechnology patents in some detail, it is worth reviewing them here:

- Member nations shall protect biotechnological inventions under their own national patent laws and that they should make their laws conform to the Directive.
- Biotechnological inventions can be patented if they involve an inventive step and are susceptible to industrial application.
- Natural varieties of animals, plants, and processes cannot be patented, but biological material that has been isolated from nature and produced by means of a technical process can be patented.
- The human body in its various stages of development and the scientific discovery of a part of the body, such as a gene, cannot be patented, although a part of the human body isolated or produced by a technical process, such as a cloned gene, may be patented even if it is nearly identical to its naturally occurring counterpart.
- Inventions whose commercialization is contrary to the public morality cannot be patented.
- It enumerates some specific types of inventions that it deems to be unpatentable based on contrariness to public morals, including processes for cloning human beings or modifying the human germ-line, and commercial or industrial uses of human embryos.

4

Arguments for DNA Patenting

Chapters 2 and 3 discussed some of the scientific and legal aspects of DNA patenting. At this point it should be apparent that there is currently a solid legal foundation for DNA patents, although lawyers and judges continue to argue about various legal issues in DNA patenting. In the remainder of the book, we will use this scientific and legal framework to consider the moral and social issues regarding DNA patents. Should DNA be patentable? If we accept DNA patenting, how should we regulate or control the patenting of DNA? These are moral questions relating to the justification of patenting laws and policies. This chapter will present moral arguments for DNA patents. Chapters 5 through 9 will address various moral arguments against DNA patents. Although this chapter will present a moral analysis of this topic, many of the premises used in various arguments and counterarguments will depend on economic, social, psychological, and biological facts and assumptions.

CONSEQUENTIALISM AND DEONTOLOGY REVISITED

Chapter 1 indicated that this book would examine consequentialist and deontological arguments for and against DNA patents.[1] Chapter 3 explored one of the most influential consequentialist arguments for intellectual property protection, utilitarianism. This theory also offers some justification for DNA patents. According to the utilitarian view, DNA patents should be legal because they are likely to lead to a greater balance of good/bad consequences for society. Although DNA patents may have some harmful effects, the good effects far outweigh the bad effects when one considers the economic, scientific, medical, and agricultural benefits of DNA patents. Of course, it is possible that one might tally all of the various effects of DNA patents and reach the opposite conclusion. Thus, one may make consequentialist arguments against DNA patents. Consequentialist arguments for or against DNA patents are forward-looking and depend heavily on empirical facts and assumptions relating to the likely effects of DNA patents on society. To evaluate these arguments, one must have a good understanding of the effects of DNA patents on science, medicine, agriculture, industry, and culture (Resnik 1997; Resnik 2001a). Chapters 6 through 9 will consider consequentialist arguments against DNA patents. Most of these arguments raise the objection that DNA patents threaten science, agriculture, medicine, and our respect for the value of human life. Chapter 6 will introduce a strategy for evaluating threats known as the Precautionary Principle.

Chapter 3 also discussed the libertarian argument for intellectual property, which takes a nonconsequentialist or deontological approach to the issue. Libertarianism is generally regarded as a deontological theory because its key terms are not defined by referring to the consequences of actions. Libertarians hold that all human beings are endowed with certain natural rights such as rights to life, liberty, and property (Nozick 1974). A libertarian argument for free speech, for example, would derive this right from an inherent right to liberty and not from an analysis of the overall good consequences of having rules in society that allow free speech. A libertarian argument for DNA patenting would focus not on the consequences of patenting rules but on the inherent right to patent, which would be based on the right to property. This book will also examine deontological arguments against DNA patents. Chapter 6 will examine the deontological argument that DNA patents are immoral because they violate human dignity (Resnik 2001b).

THE LIBERTARIAN ARGUMENT FOR DNA PATENTS

The U.S. Constitution addresses many different rights, but it does not speak of a right to patent. Patenting entered the Constitution as one of Congress' enumerated powers, similar to the power to raise taxes, spend money, coin money, declare war, or regulate interstate commerce (U.S. Constitution 1787, art. 1, sec. 8). It is Congress that has the power to grant patent rights to inventors (for limited times), which it does through the Patent Act. The Patent Act, however, does not create a federal right to patent; it merely states rules the government will follow in deciding whether to grant a patent (U.S. Patent Act 1995). Once a person obtains a patent, the patent becomes his property, and he therefore has property rights that are protected by the fourteenth Ammendment's due process clause (Resnik and De Ville 2002). If one has a patent, the government cannot take it away without due process of law. However, U.S. laws do not create a legal right to patent. Is there something like a moral right to patent?

While I think this question is worth considering, I do not regard the libertarian viewpoint as a very powerful or convincing argument for DNA patents. First, one might object to the notion of any natural rights and argue that all rights are based on moral duties. For instance, the right not to be killed is ultimately derived from the moral duties we have not to kill other human beings. If rights are derived from duties, then they have no independent moral justification and must always reflect some proper weighing of moral duties. Second, one might argue the concept of a moral right is unproductive and divisive. By casting social and political debates in terms of claims about individual rights, we frame issues and dilemmas in terms of conflicts of rights and therefore undermine our ability to reach meaningful compromises through careful deliberation. I think these are both important philosophical and political questions, but I will not pursue them here.[2]

Third, one might accept the idea there are some moral rights, or at least that moral rights play a useful role in social and political debates, but argue that property rights are not as important as other rights. Since rights frequently conflict, we must have some way of prioritizing rights and resolving conflicts (Feinberg 1973). When a property right conflicts with a right to life, the right to life takes precedence. For example, if John Doe is driving a car that goes suddenly out of control and he can either crash his car into a pedestrian, or run the car into a farmer's board fence, we would all say that he should crash his car into the fence. Although one should not damage property without good cause, in this case the pedestrian's life is more important than the farmer's fence. Other rights may also take precedence over intellectual property rights. Suppose that company Z has patented a medicine and does not intend to produce or sell the medicine but has only patented the medicine in order to gain a competitive advantage over other companies. Suppose that a country in which company Z holds a patent on the medicine now faces a public health crisis and requires access to the medicine to prevent an epidemic. In this case, one might argue that the government would be morally justified in overriding the company's patent in order to make the medicine available to people who need it. The company might still have a right to receive royalties for the medicine, but the government would be justified in using compulsory licensing to make the medicine available to the public (Resnik 2001c; Resnik and De Ville 2002).

Why are property rights less important than other rights? This is an important question I cannot explore in depth here, but I will sketch the outline of an answer. Rights should be viewed as morally or politically significant insofar as they protect the welfare, autonomy, and dignity of individual people. Rights closely connected to these values are more important than rights that are not as closely connected to these values. The right to liberty (e.g., freedom of thought, movement, and expression) is a right closely connected to these three important values because liberty is a necessary condition for exercising autonomy and for having dignity. The enjoyment of liberty also promotes individual welfare. The right to property, though it certainly has some connection to individual welfare, autonomy, and dignity, is not as closely connected to these values as other rights. The right to property is not a necessary condition for autonomy or dignity, although it does tend to promote these values as well as individual welfare. In modern societies people can vote, speak, enter contracts, hold a job, form associations, engage in political debates, create artisitic works, participate in a religious tradition, and even have children without owning property.[3] However, it would be very difficult or impossible to engage in these activities without basic liberties.

There are some additional reasons why intellectual properties do not have a close connection to individual welfare, automony, and dignity. First, many of the intellectual properties that exist today are not the products of a single individual (Resnik 1998b). Dozens of researchers may contribute to developing a genetic test, a new drug, or a genetically modified organism. A paper in genomics or biotechnology may have fifty or more authors. A database may be compiled by dozens of people and may use data that has been produced by thousands of people. Second, most research projects today have a great deal of direct or indirect government support. The government provides direct support for research through the billions of dollars that are allocated to the NIH, the National Science Foundation (NSF), and other agencies. The government provides indirect support for all research by supporting research in the basic sciences, which may be used in developing new technologies and practical applications. Since new ideas, inventions, original works, and scientific discoveries usually depend quite heavily on contributions from many individuals and the government, intellectual properties are not closely connected to individual welfare, autonomy, and dignity.[4] Indeed, these properties are often most strongly connected to the welfare and autonomy of corporations, universities, and research institutions.

These preceding points weaken the libertarian argument for DNA patents. The right to patent is more like a privilege granted by the state than a basic human right, such as the right to life or liberty. A privilege, such as driving an automobile, may be denied, restricted, limited, or regulated for worthy social purposes such as public safety and criminal justice. Thus, patent rights, though important, are far from absolute and may be compromised, restricted, or limited for legitimate social purposes. The stronger argument for DNA patents is one that appeals to the consequences of granting DNA patents instead of a purported right to patent DNA.

THE UTILITARIAN ARGUMENT FOR DNA PATENTS

The utilitarian argument is the strongest argument for DNA patents. To evaluate this argument, one must describe and understand the probable effects (both good and bad) of DNA patents. The utilitarian argument for DNA patents asserts that, all things considered, the good consequences of DNA patents outweigh the bad. There are at least four types of potentially beneficial consequences related to DNA patents: economic, scientific, medical, and agricultural.

THE ECONOMIC CONSEQUENCES OF DNA PATENTS

DNA patents have played a key role in the development of the biotechnology (biotech) industry by providing businesses with intellectual property protection. Businesses require this protection in order to protect their R & D investments. For example, R & D in pharmaceuticals is risky, time consuming, and expensive. It takes an average of ten years to test a new drug and bring it to market at an average cost of $500 million per drug. Since the life of a patent is twenty years, and it may take three years to receive a patent once it has been filed, pharmaceutical companies often have only a seven-year window to earn back their R & D investment. Only 30 percent of the drugs a pharmaceutical company develops will make it through the R & D process and be brought to the market. Only one third of those will be deemed to be successful. Thus, about 10 percent of the drugs developed net a profit (Goldhammer 2001; Resnik 2001c).

The biotech industry faces similar economic constraints and pressures, since biotechnological products also cost millions of dollars and many years to develop and implement (Biotechnology Industry Organization 2000). For example, Celera Genomics has been in existence since 1997. From 1997–2001, the company spent about $200 million per year in R & D but had not yet turned a profit. In the fiscal year ending in 2001, Celera had $186 in losses (www.Celera.com). The industry could probably survive with less patent protection than it currently has, but it depends quite heavily on patents. Although low tech industries, such as the food or travel industry, do not depend heavily on patents, high tech industries, such the biotechnology and pharmaceutical industries, depend heavily on patents (Scott 2000; Woollett and Hammond 1999; Tribble 1998).

The biotech industry has developed patents on many different biological materials and processes, including cell lines, proteins, genetically modified organisms, and PCR. Because DNA is such a key molecule in biology, patents on DNA play a central role in the overall intellectual property scheme in biotechnology. The top DNA patent holders in 1999 included: Incyte Pharmacueticals—356 patents; University of California—265 patents; SmithKline Beecham—197 patents; Genentech—175 patents; Eli Lilly—145 patents; Novo Nordisk—142 patents; Chiron—129 patents; American Home Products—117 patents; Isis Pharmaceuticals—108 patents; Massachusetts General Hospital—108 patents; Human Genome Sciences—

104 patents; University of Texas—103 patents; Institut Pasteur—101 patents (Biotechnology Industry Organization 2000).

Since the late 1980s, the biotech industry has expanded at a rapid pace as new companies have started up and existing pharmaceutical companies have expanded their operations into the biotechnology sector. According to industry estimates, the industry doubled between 1993 and 1999 (Biotechnology Industry Organization 2001a). The industry includes new companies, such as Celera Genomics, Amgen, Applied Biosystems, GeneOp, Strategene, Biolabs, and Human Genomic Sciences, as well as older companies, such as Incyte Pharmaceuticals, Glaxo Welcome, Genentech, and Monsanto. Biotech companies sponsor research in basic sciences like genomics, proteomics, molecular biology, and bioinformatics, as well as research in applied fields, such as gene therapy, genetic testing, agricultural biotechnology, cancer genetics, pharmacogenomics, and drug discovery. Although some biotech companies are traded on the New York Stock Exchange, most of them are traded on the NASDAQ. Even critics of the industry, such as Rifkin (1998), recognize that the twenty-first century will be the biotech century.

The biotech industry invests billions of dollars each year in R & D and has created thousands of high-paying, white-collar jobs for scientists and technicians. According to industry estimates (Biotechnology Industry Organization 2001a), the industry spent $11 billion on R & D in 1999 compared to an estimated $26 billion spent by the pharmaceutical industry (Resnik 2001c). In the United States, private industry invests over $100 billion per year in R & D, more than half of all total R & D outlays (Resnik 2001a). The biotech industry accounted for 437,000 U.S. jobs in 1999, including 151,000 direct jobs and 287,000 indirect jobs. The industry had $47 billion in revenues in 1999, including $20 billion in direct revenues and $27 billion in indirect revenues. The industry paid $11 billion in taxes to federal, state, and local governments (Biotechnology Industry Organization 2001a).

The biotech industry has had measurable effects on local economies in the states of Maryland, Massachusetts, Pennsylvania, Delaware, Michigan, California, and North Carolina. The industry has also had a significant impact on the global economy (Enriquez 1998). Countries around the globe, including England, Germany, Israel, Japan, India, Australia, Indonesia, Brazil, and China, are courting biotech companies in order obtain the economic benefits related to business expansion. In the United States, fourteen different states have developed bioscience strategic plans, twelve have staff specialists who deal with bioscience, and twenty-nine states have collected data on the biotech industry. Several different cities, including St. Louis, Pittsburgh, San Antonio, Cincinnati, and Columbus are engaged in efforts to develop the bioscience sector (Biotechnology Industry Organization 2001b).

No utilitarian argument is complete without addressing costs, risks, and harms. DNA patents have potential economic harms related to increased costs for new technologies in science, medicine, agriculture, and other fields where DNA patents play an important role in research and product development. For example, the initial cost of Myriad Genetics breast cancer test was $2400, although the company has licensed several laboratories and the National Cancer Institute to conduct the test for less money (Foubister 2000). Many geneticists, oncologists, and other physicians have complained about the high cost of this genetic test, and they are wary of the proliferation of other expensive, privately owned tests as biotech companies develop more genetic tests (Reynolds 2000). Amgen's drug, Epogen, is also very expensive: Medicare spends $1 billion per year on the drug. Amgen has increased the price of the drug two years in a row. Many health care advocates, including the National Renal Administrators Association, have complained about the price of Epogen (Goozner 2000). These and other cases could forebode a trend toward high-tech, high-cost medicine. (We will consider this issue again in chapter 8.)

In response to the cost argument, one might argue that biotech medicine is no different from other types of medicine: any type of medicine can be expensive. The key factors contributing to the cost of any medical therapy, test, or procedure are the degree of patent protection and the degree of market competition. Prices are highest when patent protection is great and competition is low because a company can take advantage of its monopolistic control of the market. In order to whittle away this control, competitors can develop similar inventions and try to enter the market. For example, to compete with Amgen, another company could develop a different type of drug that stimulates red blood cell production, which would force Amgen to lower its cost in order to attract customers. If Amgen manages to corner the market on products for stimulating red blood cell production by obtaining many different patents or patents with broad scope, then it may be able to maintain its monopoly position. Amgen, like any other company, will use a variety of patenting strategies to recoup its R & D and marketing costs, since once its patents expire, the monopoly on the inventions ends, and prices drop dramatically. However, it is worth noting that the United States and other countries have antitrust laws for dealing with monopolies and unfair competition. It would be prudent to apply these laws to the biotech industry in order to deal with companies that try to thwart competition unfairly. (We will discuss this point again in chapters 8 and 10.)

Those who object to the high costs associated with medical therapies and tests related to DNA patents also argue that it is unfair for private companies to exact such high profits from basic research that was sponsored by the government. For example, the NIH sponsored much of the preliminary research on the sequences of BRCA1 and BRCA2 mutations. Myriad Genetics also funded some basic research on BRCA1 and BRCA2, but the company invested most of its capital and labor in developing a test based on this research (Foubister 2000; Reynolds 2000). The NIH also funded much of the preliminary work relating to the erythropoietin gene, and Amgen developed methods for making Epogen from this research.

The Bayh-Dole Act, which was passed in 1980, helped pave the way for the transfer of technology from the public to the private sector by allowing companies to patent inventions based on data and research sponsored by the government (Dreyfuss 2000). Before Bayh-Dole, it was much more difficult for companies who wanted to develop a product from government research to obtain intellectual property rights. The rationale for the Bayh-Dole Act was to allow the government to sponsor basic research and to encourage companies to make sufficient capital investments to develop products and bring them to market. It still takes a great deal of additional funding to develop a product and bring it to the market even when the basic research leading to the invention has been conducted (Goldhammer 2001).

Many people object to Bayh-Dole, claiming it is nothing more than a form of corporate welfare, since it allows private companies to profiteer from public investments in research and. (Goozner 2000; Rifkin 1998). Supporters of Bayh-Dole argue that the public benefits from private efforts to develop products and bring them to the market, since many of these inventions would not be developed and marketed without appropriate incentives for industry (Dreyfuss 2000). These critiques of the Bayh-Dole Act extend way beyond the issue of DNA patents, since companies would still be able to profit from Bayh-Dole even without DNA patents. Thus, for the purposes of the book, I will consider this important economic argument, but I will not hold that it is a decisive argument against DNA patents, since it applies to all technologies developed from basic research sponsored by the government.

Finally, those who object to the high costs of products and services related to DNA patents also argue that these costs can have a negative impact on scientific research or access to care for patients (Andrews and Nelkin 2000). There are replies to these arguments as well, but I will not discuss them until chapters 7 and 8.

To get a complete understanding of the economic costs and benefits of DNA patents, one must also come to terms with the scientific, medical, and agricultural consequences of DNA patents, since these consequences also have economic effects (Knoppers, Hirtle, and Glass 1999). Since I will examine these consequences in more depth in chapters 7, 8, and 9, my discussion of the economic effects of DNA patents will not be complete until the end of the book.

THE SCIENTIFIC, MEDICAL, AND AGRICULTURAL CONSEQUENCES OF DNA PATENTS

DNA patents benefit science in several ways. As we saw above, DNA patents have provided industry with an incentive to invest billions of dollars per year in many different areas of basic and applied research. Without this huge influx of private money, many worthy research projects would not be done or would have to compete with other projects for government funds. Although the U.S. government funds more than $30 billion per year in biomedical research through the NIH, it does not have enough funds to make up for the private sector's contributions (Malakoff and Marshall 1999; Resnik 1996). For better or worse, modern science requires a great deal of money, which does not grow on trees. In order to conduct research, someone must pay the bills. Private investment plays a key role in promoting research and innovation in biotechnology, genomics, and numerous other sciences. Many scientists now either work for the private sector or have private contracts to conduct research for industry. Many scientists, such as Celera's Craig Venter, have also started their own biotech companies based on their work in academic research. Although the amount of research conducted in the fields of genomics, genetics, and molecular biology did not double during the 1990s, it rose at a steady pace (Resnik 2001a). (I present some additional data relating to this point in chapter 7.)

Besides helping to pump more money into science, DNA patents have helped to accelerate the pace of discovery by encouraging healthy competition among researchers from the public and private sectors. For example, the race to sequence the human genome involved a competition between the NIH-led public effort and Celera Genomics. During this race, the two sides shared data and techniques even though they had different strategies and goals. As a result of this competition, the HGP was completed in 2001, several years ahead of schedule (see chapter 1). Although there were many rivalries and conflicts during this race, the two sides eventually agreed on some data release plans and policies, and both copies of the genome are now publicly available (Roberts 2001; Marshall 1999b, 2000b,d,e).

By encouraging private investments in R & D, DNA patents have had some benefits for medicine. Although many of the medical applications of genomics and biotechnology are still years away, in the last decade genomics and biotechnology have contributed to the discoveries of new drugs, genetic tests, and gene therapies. As noted in chapter 2, Amgen used DNA patents to protect its intellectual property interest in Epogen, a drug now widely administered to patients suffering from anemia. Myriad Genetics developed a genetic test, which is now used to test women for breast cancer mutations. Companies such as Avigen, Transkaroytic Therapies, Imperial Cancer Research Technology Limited, Medigen Incorporated, and Cogent Neuroscience have developed and patented methods for use in gene therapy (Patent and Technology Office 2002). However, it should be stressed that the medical benefits of DNA patents are mostly expected or predicted benefits at this point in time, because it will still take many years for the biotechnology revolution to have its full effects on medicine.

DNA patents have had similar effects on agriculture and have helped to stimulate the growth of agricultural biotechnology. Companies have used genetic engineering technologies to produce genetically modified (GM) foods, crops, and animals. Agricultural biotechnology has improved the productivity efficiency of farming and food production (U.S. Department of Agriculture 2001). As a result of incentives to invest in biotechnology, companies have sponsored billions of dollars of agricultural R & D and have sequenced the genomes of species with agricultural significance, such as rice and corn.

On the other hand, DNA patents may also have some potential negative effects on science, medicine, and agriculture related to licensing, secrecy, conflicts of interest, monopolistic control, exploitation, and other aspects of the commercialization of science, medicine, and agriculture. I will examine the potential adverse consequences of DNA patents in chapters 7, 8, and 9.

CONCLUSION

This chapter has considered two arguments for DNA patents, a libertarian argument and a utilitarian one. While the libertarian argument is important, it does not prove that there is an absolute right to patent DNA: patenting DNA is more like a privilege than a right. Since society may limit or override patent rights for important social purposes, utilitarianism provides us with a better argument for DNA patents. DNA patenting laws and policies can be morally justified insofar as they promote the greatest balance of good/bad consequences for society. This chapter provided the reader with a sketch of this argument, and the rest of the book will fill in more of the important details related to the scientific, medical, and agricultural consequences of DNA patents in chapters 7, 8, and 9.

No utilititarian argument is complete without examing all of the potential consquences. So far, this chapter has mentioned the economic, scientific, medical, and agricultural consequences of DNA patenting but it has not addressed the social consequences of patenting. How might DNA patents affect social values, practices, and beliefs? The next two chapters, chapters 5 and 6, will address some of the potential negative cultural consequences of DNA patenting and also examine deontological arguments against DNA patents.

5

Patenting Nature?

The previous chapter introduced the utilitarian argument for DNA patents, which chapters 7 through 9 will explore in more depth. This chapter and the following one will consider some deontological arguments against DNA patents as well as argument that address some of the negative social consequences of DNA patents. Deontological arguments assert that the very practice of DNA patenting is in some way inherently wrong, regardless of its consequences; the wrongness of DNA patenting lies in the act itself. This chapter will address the argument that DNA patents are immoral because they are patents on nature.

No Patents on Nature?

Ever since the courts recognized patents on biological materials, opponents have argued against these patents on the grounds that they are immoral patents on nature. As we saw in the *Chakrabarty* case, one of the fundamental issues in biotechnology patenting is whether living things are products of nature or products of human ingenuity (Resnik 2001a; Sagoff 1999). The U.S. Supreme Court decided that living things can be patented if they are the result of (or caused by) human ingenuity. The courts and the patent agencies have decided that DNA in its natural state cannot be patented; but also that DNA that has been isolated and purified or in some significant way modified, can be patented, because it results from human ingenuity. Human labor and effort can create DNA in a form that does not exist in a natural state (Doll 1998).

Most people who oppose DNA patents are not arguing against the patent system per se, since they recognize the distinction between products of nature and products of human ingenuity (Evans 1999). Many opponents of DNA patents argue, however, that all DNA, whether created in the wild or in the lab, is a product of nature, not a product of human ingenuity, and therefore should not be patented under any circumstances (Rifkin 1998; Andrews and Nelkin 2001; Shiva 1996; Kimbrell 1997). The press release from the Joint Appeal against Human and Animal Patenting states: "We believe that humans and animals are creations of God, not humans, and as such should not be patented as human inventions" (Joint Appeal 1999, 169). Although this statement mentions only humans and animals, the rest of the declaration implies that a prohibition on patenting should apply to all biological materials (Peters 1997). The Council for Responsible Genetics (2000, 1), a nonprofit group concerned with ethical and social issues in genetics, has adopted a Genetic Bill of Rights, which states: "All people have the right to a world in which living organisms cannot be patented." These declarations hold DNA should not be patented, regardless of the consequences for science, medicine, or agriculture. The patenting of DNA is simply and straightforwardly immoral, according to these critics of DNA patenting. These declarations also do not recognize a distinction between natural and nonnatural DNA. All DNA patenting is wrong.

On the other hand, many international agreements and declarations take a view similar to the ruling in *Chakrabarty:* they draw a distinction between natural and non-natural DNA and declare that biological materials in their natural state should not be patented. As noted in chapter 3, a 1997 UNESCO declaration condemns profiteering from the human genome in its natural state. According to Article 4 of this declaration, "The human genome in its natural state shall not give rise to financial gains" (United Nations Educational, Scientific and Cultural Organization 1997). Chapter 3 also noted that the EPC holds that biological materials and processes can be patented if they result from an inventive step (European Patent Convention 1998). According to the European Parliament and Council (1998), natural varieties of plants, animals, and processes cannot be patented, but biological materials that have been isolated from nature and purified can be patented.

Clearly, any coherent discussion of DNA patenting must resolve the following two issues: (1) Is DNA a product of nature or a product of human ingenuity? (2) Why should there not be patents on products of nature? I will begin with a discussion of the second question.

THE UTILITARIAN ARGUMENT
AGAINST PATENTING NATURE

There are several arguments against patenting products of nature. The argument, which I consider to be the most convincing one, is essentially utilitarian (or consequentialist) in form. The three other arguments are deontological. According to the utilitarian argument, we should not allow products of nature to be patented because this will stifle scientific innovation and discovery. Imagine the detrimental consequences of allowing scientists to patent newly discovered species, mathematical formulas, or natural phenomena, such as fission. Private ownership and control of natural phenomena would have negative effects on the progress of science and technology. This argument against patents on nature coheres with the utilitarian perspective on intellectual property in general: intellectual property laws and policies, including definitions of what is patentable, should promote social goals (Miller and Davis 2000; Resnik 2001a). The utilitarian argument against patents on nature does not constitute a deontological argument against patenting nature, since the strength and scope of the argument depends on some assessment of the consequences of various laws and policies. The utilitarian argument also need not recognize any absolute demarcations between products of nature and products of human ingenuity: the important policy question, from a utilitarian perspective, is how to understand the consequences of various policies and how to develop interpretations that distinguish between products of nature and products of human ingenuity. I will return to these concerns later on in this chapter.

PATENTING GOD'S CREATION

The other three arguments against patenting nature are deontological in character. First, one might argue that products of nature are things that have been created by God, not by human beings (Mitchell 1999; Land and Mitchell 1996). The Bible's book of Genesis 1:1 states: "In the beginning, God created the heaven and the earth." It also declares that God made animals, plants, all living things, and human beings. To patent a product of nature is to play God, or at least to exert undeserved and improper authorship and dominion. According to Mitchell, "by claiming exclusionary property rights in a genetically-altered composition of matter, the human manipulator is assuming a place which belongs alone to God, the divine Artificer" (Mitchell 1999, 176). If we translate this religious argument into the language of patent law, we could say that God's acts of design or creation undermine any claims to the originality of the purported invention; God is the original creater, not human beings.

There are several reasons why one might reject appeals to religion in ethical and policy debates. First, religious beliefs cannot serve as basic premises in legal arguments in the United States due to the separation of church and state (Barron and Dienes 1999). The claim that a patent should be illegal because it is against God's will or violates God's dominion would have little force in any federal court. Second, one might argue that we should not base public policy arguments on appeals to religion because religious beliefs and doctrines are highly diverse and often irreconcilable. Public policy arguments should appeal to broad moral and political principles that can emerge from a broad consensus (Rawls 1993). On the other hand, religious beliefs often have a great deal of persuasive power in shaping and influencing moral, legal, and public policy debates and they therefore should not be excluded from the discussion. Indeed, the National Bioethics Advisory Commission (NBAC) made a concerted effort to include religious viewpoints in its reports on human cloning and stem cell research, due to the importance of religion in American public life (1998). Thus, I believe it is important to consider this religious argument against DNA patenting.

For those who hold religious beliefs (including this author), patenting nature could be viewed as a form of hubris (or arrogance) because it would assert authorship and exclusive rights over something that was not created by man. God, not man, has the right of authorship and exclusive control over nature. This view makes sense, from a theological perspective. Human beings did not make nature. All of the world's main religions have accounts of the creation of the world in which God makes or designs the world, including man, and gives man stewardship of nature. As stewards of nature, human beings have obligations to take care of nature and share natural resources (Mitchell 1999). God created, invented, and owns everything, according to this view.

This view offers an important perspective on divine creation and human participation in the world. However, one still must ask how human beings should act as stewards of the world. Shouldn't human beings still divide property among themselves even if, in some ultimate sense, God owns everything? Shouldn't human beings take credit for inventions and original works even if, in some ultimate sense, God is the author of everything? Even if one believes that God is the creator of the world, one must still come to terms with practical issues relating to human responsibility, agency, and ownership (Peters 1997; Cole-Turner 1999). If I take some clay and mold a clay statue from the clay, we can say that God has made the clay and He has made me, but should we say that he made the statue? Am I not the author of the statue? Or suppose I take the clay and make it into a useful invention, such as a candleholder. Should we say that God invented the candleholder? Unless we take the extreme position that human beings have no authorship in making artifacts and have no exclusive rights (among human beings) to use, copy, make, or market those artifacts, then we must adopt some type of distinction between God-made things, which the legal system calls "products of nature," and human-made things, which the legal system calls "products of human ingenuity." Indeed, many churches have wrestled with this very issue of human versus divine creation and the concept of cocreation (Cole-Turner 1997).

If we accept some kind of distinction between God-made and human-made things, then we need some way of classifying things as God-made or human-made. Deciding these sorts of issues takes the debate out of the purely theological realm because deciding how to interpret and apply a definition of human-made is more of a practical matter than a theological one. Even if one operates within this theological framework, one must still face familiar questions, such as a whether a house, a mouse, or a strand of DNA is or is not a human artifact. Thus, while this theological perspective can give us an answer as to why it is wrong to patent nature, it still leaves open the question of what counts as natural and therefore unpatentable.

DNA AS OUR COMMON HERITAGE

The second deontological argument asserts that it is immoral to patent human DNA because the human genome is our common heritage: DNA belongs to everyone and should not be privately owned or commercialized (Sturges 1997). Many different organizations have championed the idea that the human genome is our common heritage, including the Human Genome Organization (2000), the Council on Responsible Genetics, and the World Health Organization. A declaration by UNESCO says: "The human genome underlies that fundamental unity of all members of the human family . . . in a symbolic sense, it (the human genome) is the heritage of humanity" (United Nations Educational, Scientific and Cultural Organization 1997, 1).

The common heritage argument against patenting natural things makes use of the idea of *res communis* (or common property), which is property that is owned by all, or is at least available to all. For many years, the oceans were regarded as *res nullis* (or things owned by no one), until they were declared the common heritage of mankind *(res communis)* by the Law of the Sea Convention (Ossario 1999). Other types of *res communis* include the moon, Antarctica, and the atmosphere. Another concept of common property is public domain, which would include properties owned by the government, such as parks or streets, which are open to all. As noted on chapter 3, intellectual property law makes extensive use of the concept of public domain, which includes things that are not the subject of copyrights or patents, such as facts and laws of nature, as well things whose copyright or patent has expired. In economic theory, the idea of a public good is a good that benefits society but must be provided by the government because individuals lack the incentives or the ability to provide the good. Public goods include streets, clean water, public schools, and the police (Samuelson 1980). The key idea uniting all of these different concepts is that there are some domains that should not be privately controlled or owned.

There are different ways of justifying common properties. As we saw above, one might take a utilitarian view that society benefits by forbidding private ownership of some things, such as the oceans, the air, streets, or certain types of scientific or technical knowledge. One might make an economic argument for public goods based on a cost/benefit analysis as well. In both cases, one appeals to the consequences of public ownership and control. Are there deontological arguments for *res communis?*

The common heritage argument can be construed as a deontological argument against private ownership if one argues that some things literally belong to everyone. The argument might go like this. Suppose that a farmer owns 400 acres of land and that in his last will and testament he leaves it to his two sons and two daughters, 100 acres apiece, with the provision that they each have access to a river that runs through the property. The will could even have a stipulation that all subsequent generations of property owners shall have access to the river. The river is their common property. If one of the sons or daughters attempts to assert exclusive possession or control over the river, for whatever reason, then he or she has violated the property rights held in common (i.e., their common heritage).

Is the human genome literally our common heritage? As noted in chapter 2, human beings share 98 percent of the DNA with chimpanzees and a great percentage of their DNA with other species, including flies and yeast. Only 2 percent of the human genome is exclusively human in character (i.e., occurs only in human beings).[1] So, only 2 percent of the genome is actually our common heritage; the other 98 percent belongs to other species on the planet, assuming that this idea makes any sense. Moreover, approximately 0.1 percent of the exclusively human DNA consists of SNPs, which vary from person to person. A tenth of a percent may sound like a small number, but this amount of genetic variation accounts for much of the genetically caused phenotypic variation we find in our species. Putting all of these facts together, it does not make a great deal of sense to say that the human genome is literally our common heritage because most of the genome is not exclusively human, and the parts that are exhibit significant variation. It may be our common heritage in a symbolic or metaphorical sense, but it is not literally our common heritage. The genome as common heritage is an empirical fiction (Juengst 1998). Therefore, DNA patenting does not literally violate a property right in a common genetic heritage (Ossario 1999). Incidentally, this result should come as no great surprise because the same type of argument also applies to the oceans: the oceans are not literally our common heritage.

If the common heritage idea should be understood only symbolically, then it loses its force as a deontological argument against ownership of DNA. Symbols may have moral value, of course, but their value must be understood with reference to the people who use the symbols. Thus, an American flag has value not for its own sake but for the sake of people who value this symbol (e.g., U.S. citizens). The act of burning a flag is not inherently immoral but it is immoral insofar as it has harmful consequences: burning the flag is wrong because people find flag burning to be offensive. Thus, the common heritage

argument, properly construed, raises consequentialist but not deontological
objections to the ownership of DNA. To assess this argument, we must therefore assess the consequences of ownership and control of DNA, and this con-
squentialist perspective raises familiar questions about the benefits versus
harms of private ownership and control of DNA.

By claiming that the common heritage argument is not a convincing
deontological objection against DNA patenting, I do not mean to imply that
it does not raise important points, since it does. Indeed, the argument focuses
our attention on the important question of how we should be good stewards
of our natural resources, including genetic resources, and how we should share
those resources among all the people in the world. The common heritage idea
does not prohibit commercialization of the human genome any more than it
prohibits commercialization of the moon or the seas (Ossario 1999). However, it does compel us to address important questions concerning global and
intergenerational distributive justice (Looney 1994; Cahil 2001). It is likely
that the patenting of DNA will benefit industrialized nations as well as
biotech companies. It will also benefit current generations, as well as many
other nations and people living in the present and the future. If we are concerned about the symbolic value of DNA as *res communis,* then we should
address questions concerning the impact of DNA patenting on the distribution of benefits and burdens in society.

Theories of distributive justice address the just distribution of benefits
and burdens in society. There are three basic approaches to deciding how to
distribute benefits and burdens justly: the libertarian approach, which holds
that we should distribute benefits and burdens according to rules that protect
basic rights and fair acquisition and transfer of property (Nozick 1974); the
egalitarian approach, which holds that we should distribute benefits and burdens by following rules that protect basic rights and promote equality of
opportunity (Rawls 1971), and the utilitarian approach, which holds that we
should distribute benefits and burdens by following rules that maximize utility (Mill [1861] 1979). I will not attempt to compare, contrast, or critique
these theories in this book. However, I would like to point out that only the
egalitarian and utilitarian approaches would raise significant concerns about
injustices relating to the commercialization of DNA. On the libertarian view,
the patenting of DNA raises no significant problems for distributive justice,
provided that inventors and companies follow the legal rules for patenting,
assigning patents, licensing, etc.

The utilitarian view is simply the application of utilitarian theory to problems of justice. A utilitarian would say that the commercialization of DNA is
just, provided that it results in a distribution of benefits and burdens that maximizes social utility. The utilitarian would acknowledge that there might be an
unequal distribution of benefits and burdens from the genome: some people
may benefit much more than other people will benefit. The utilitarian would

argue that differences in the distribution of the benefits from DNA patenting are just if they maximize utility.

John Rawls, the most influential egalitarian theorist, would also allow an unequal distribution of benefits from DNA patenting, provided that this distribution benefits everyone, including the least advantaged members of society. According to one of Rawls' principles of justice, known as the difference principle, differences in the distribution of the benefits in society can be justified if (1) they benefit everyone in society, especially the least advantaged members, and (2) they support and do not undermine equality of opportunity (Rawls 1971). When Rawls defended the difference principle, he was thinking about the distribution of wealth in society as well as various rules that tend to redistribute wealth, such as progressive income taxes, estate taxes, and government welfare programs designed to fight poverty, ignorance and disease. But one could apply Rawls' ideas to address the distribution of wealth (and other benefits) related to DNA patenting. Many people have argued that DNA patenting will benefit a few people at the expense of millions. Biotech companies in industrialized nations will benefit from patenting, but poor people around the world will not benefit. DNA patenting will only add to the growing gap between rich and poor around the world (Rifkin 1998; Kimbrell 1997; Andrews and Nelkin 2001). Critics of patenting also argue that it drives up the price of essential medicines and makes them unaffordable to people in the developing world. For example, genetically engineered clotting factors, which are used to treat hemophilia, can cost thousands of dollars per month. It is estimated that most hemophiliacs will exceed their lifetime health insurance caps after the second decade of life. To prevent these injustices, some would argue, we should radically restrict DNA patenting or forbid it altogether.

These arguments contain several flaws. The high price of medications does not itself demonstrate that the patent system is unjust because these medications might not even be available, unless a private company had invested money to develop them. The important question to ask is not, "are benefits and burdens distributed unequally?" but "Does the unequal distribution of benefits and burdens make everyone better off?" While it is regretful that hemophiliacs must pay such a high price for synthetic clotting factors, one should consider the alternatives. Before synthetic clotting factors became available, hemophiliacs had to use natural clotting factors extracted from human blood. These clotting factors not only cost a great deal of money to produce and were in short supply, but they also carried a high risk of HIV infection. As a result, many hemophiliacs died of AIDS. Some day in the not too distant future patents will expire on synthetic clotting factors, which will dramatically lower their price. In the short-term, it seems highly unfair that people do not have access to these medications due to their prohibitive cost. However, in the long run, the cost will go down and people will have greater access. Of course, governments could eliminate these initial high costs by sponsoring biotechnology research and development so that patients would not need to depend on private investment to obtain access to new medications. But does any government, including the U.S. government, have $70 billion to spare on more biomedical R & D? Unless governments find a way to pay for all biomedical research and development, private companies will have to fund it, and private companies will only fund R & D if they have the expectation of a reasonable return on their investments, which the patent system helps to ensure (Resnik 2001c).

If one focuses on the long run consequences of biotechnology patenting instead of the short-term inequities, it is highly likely that DNA patenting will benefit many people and many nations, including rich people in rich nations, poor people in rich nations, rich people in poor nations and even poor people in poor nations. These benefits could occur from applications of genomics research in medicine and agriculture, sponsored by private companies. Other chapters in this book will address questions related to the medical and agriculture benefits of DNA patents. To satisfy the demands of distributive justice, countries should take steps to ensure that patents promote benefit-sharing (i.e., that the benefits of research will accrue to society, not just to private companies). Many of the rules relating to patents promote benefit-sharing by striking a balance between public access and private control. Thus, a proper resolution to many of the legal issues discussed so far, such as distinguishing between products of nature and products of human ingenuity, meeting the utility criteria, delineating the scope of patents, developing a research exemption, and even setting the proper length of the term of a patent, help promote benefit-sharing.

If we look beyond the dispute about DNA patents and consider other technologies, one finds that the overall positive effects of patenting benefit poor people in the world because the products developed from patented inventions benefit poor and rich people alike. Granted, new technologies are usually expensive at first and may produce short-term inequities. It may take a while for benefits to trickle down from rich to poor, but over time, the benefits of technological development eventually touch everyone in society. New technologies eventually become inexpensive and highly accessible. For example, an electronic calculator cost about $300 in 1972. The same device costs less than $5 today. The following important inventions in the twentieth century were protected by patents at one time but are now accessible and affordable for many people: personal computers, automobiles, antibiotics, anti-inflammatory drugs, pacemakers, purified vitamins, radios, phones, and microwave ovens. As long as society establishes a patent system that provides an appropriate balance of public access versus private control, it is likely that the distribution of the benefits of patenting will be just.

Issues of distributive justice also arise with respect to future generations: will DNA patents benefits people in the current generation at the expense of future generations? Before one considers issues of justice with respect to future generations, one should ask the logically prior question, "Do we have any moral duties to future generations?" Duties to future generations raise a variety of metaphysical and epistemological questions. How can we have a duty to some-one who does not yet exist and may never exist? How can we have a duty to someone if we do not know who they are, or what they want or need? For the purposes of this book I will assume that these philosophical problems can be solved and that we do have moral duties to future generations (see Baier 1984; Shrader-Frechette 1991). However, since future generations are abstract and remote, our duties to current generations take precedence over our duties to future generations: if I can use a dose of medicine to save a child who is right before me, I should use that dose now, instead of saving it to use on some child who might be born in the future and might not need the medicine:

The basic moral duty that we have to future generations is a duty of stew-ardship. We should take care of the earth's resources to ensure that they are available to future generations (Rolston 1994). By fulfilling the duty of stew-ardship, we can satisfy the demands of intergenerational justice. Many coun-tries have enacted environmental protection laws in order to ensure that cur-rent and future generations have access to clean air, land, and water. Since the 1960s, the United States has a passed a slew of environmental legislation, such as the Clean Air Act, Clean Water Act, Endangered Species Act, and the Waste Disposal Act. Although no country, to my knowledge, has passed any legislation seeking to protect genetic resources, it is conceivable that a nation might enact such legislation at some future time. Genetic diversity is proba-bly the most important genetic resource, since genetic diversity plays an essen-tial role in evolution and ecosystem stability. Laws that protect endangered species indirectly protect genetic diversity, since each species makes its own, unique contribution to the world's genetic diversity. Besides preserving native species, one might also preserve genetic diversity by storing genetic samples from species that are about to go extinct. These samples would represent a reservoir of genetic diversity for future use. We will return to questions about genetic diversity when we examine agricultural biotechnology in chapter 9. Critics of DNA patenting argue that it can threaten genetic diversity by pro-moting agricultural practices, such as cloning, that undermine diversity.

COMMODIFYING NATURE

The last deontological argument I will consider in this section is the argument that patenting nature is immoral because patenting is a form of the commod-ification of nature that violates the dignity, value, or sanctity of the natural world. In chapter 6, we will address these concerns as they relate to the patent-ing of human beings or human body parts, but one might also develop this argument as an objection to the patenting of any natural thing, including a protein, a plant, an animal, and so forth (Hanson 1997, 1999). As we saw earlier, the Joint Appeal raised this type of objection against DNA patenting: DNA is part of nature; patenting nature violates the sanctity of nature; hence, DNA patenting violates the sanctity of nature (Joint Appeal 1995). Others, such as Shiva (1996), Kimbrell (1997), and Rifkin (1998) have also developed this argument against DNA patenting.

If we take this argument at face value, then it would also imply that other types of property control over nature are also immoral because they commodify also nature. As we saw in chapter 3, patenting is one type of commodification, but there are many other forms, such as renting, buying, copyrighting, selling, and so on. If patenting an animal violates the dignity of that animal, then certainly selling the animal violates its dignity. Currently, most people in the world accept various forms of ownership of nature, ranging from ownership of land and livestock to ownership of crops and seeds. Unless one takes the radical view that these widely accepted practices are also immoral, then the argument from the dignity of nature does not make a great deal of sense.[2] On the other hand, one could agree that some forms of the ownership of nature are morally acceptable but object to patenting as particularly degrading and offensive. But why focus on patenting? Perhaps what is most offensive about patenting is that it implies that some human being is the inventor or creator of something that is purported to be natural. But this suggestion brings us back to our earlier question about whether the patented item really is a product of nature.

SYNOPSIS OF ARGUMENTS AGAINST PATENTING NATURE

To summarize our discussion in the preceding three sections, there are several deontological arguments against patenting nature. In my judgment, the theological argument is the most coherent deontological argument against patenting nature, since the other two arguments suffer from empirical and analytical difficulties. The main difficulty with the theological argument is that it makes theological assumptions that many people may not accept. Although the common heritage argument raises some important points about DNA patenting, it does not prove that DNA patenting is inherently immoral. It proves that DNA patenting is morally acceptable, provided that we address questions of global and intergenerational justice. The theological argument, on the other hand, implies that DNA patenting is inherently wrong: patenting nature is immoral because it arrogantly asserts that human beings have created or invented something that was made by God. Since many people will not be swayed by the theological argument, the best argument against patenting nature is the utilitarian argument that focuses on the adverse consequences of patenting nature. From this perspective, patenting nature is immoral because the harmful effects of patenting outweigh the beneficial effects. In any case, to apply any one of these arguments to patenting laws or policies, one must appeal to some kind of distinction between a product of nature and a product of human ingenuity. Is DNA a product of nature or a product of human ingenuity? It is time to confront this question directly.

PRODUCTS OF NATURE VERSUS PRODUCTS OF HUMAN INGENUITY

As we saw in Chapter 3, the ruling in *Charkrabarty* held that products of nature cannot be patented, but that products of human ingenuity can be patented. We have also seen that various polices, laws, and declarations forbid the patenting of natural DNA while allowing the patenting of isolated and purified DNA. Under U.S. patent laws, several categories of things have been viewed as products of nature. These include:

1. Laws of nature, such as Einstein's equation $E = MC2$.
2. Natural processes, such as boiling, digestion, or flight.
3. Natural elements or compounds, such as oxygen, water, or sugar.
4. Natural species, such as whales, bamboo, or mushrooms.
5. Abstract ideas, such as mathematical equations and algorithms, such as the quadratic formula or Bayes' theorem.

Where does DNA fit into this scheme? As we saw in chapter 2, the PTO's view of DNA is that it is a chemical that occurs in living things and encodes information for making proteins. The PTO has holds that when one isolates and purifies (or in some other ways alters) this natural compound, that one makes a compound that can be patented. Opponents of DNA patenting, however, argue that these processes do not transform DNA into an artificial compound that can be patented. How can we clarify this dispute?

For starters, it is important to realize that this type of dispute is not unique to DNA patenting, since there are many other cases of patent disputes related to subject matter. For example, in the information technology industry, there has been a dispute over the patenting of algorithms used in computer software (*Parker v Flook* 1978). Some have argued that algorithms cannot be patented because an algorithm is nothing more than a set of mathematical formulas and procedures. Others have argued that an algorithm can be patented if it is reduced to a tangible form (i.e., if it is instantiated in a machine). The algorithm itself cannot be patented, but an application of the algorithm can be patented. In these types of disputes, we must draw a line between ideas and applications of ideas. These controversies arise because we have trouble determining whether something is application of an idea.

Prior to DNA patenting, disputes arose concerning other chemical compounds that had been isolated and purified. In *Merck & Co. v. Olin Mathieson Chemical Corporation* (1958), the U.S. Fourth Circuit court addressed the patentability of naturally occurring vitamin B_{12}. The vitamin occurs in nature, but the court held that an isolated and purified form of the vitamin is patentable. Although vitamin B_{12} is not as large or important a molecule as DNA, it stands in a similar relationship to patent law and policy: can vitamin B_{12} be patented? To decide this question, the courts have, in effect, drawn a distinction between natural vitamin B_{12} and artificial vitamin B_{12}. Artificial vitamin B_{12} is patentable; natural vitamin B_{12} is not. As *Chakrabarty* did two decades later, the court drew an implicit distinction between natural bacteria and artificial bacteria, declaring only the latter patentable. A similar type of dispute may arise relating to a natural chemical process, such as phosphorescence. Phosphorescence, as a natural process, cannot be patented but one may patent an application of this natural process (e.g., using phosphorescence to determine whether a protein has bound itself to another compound).

From these examples, we can see that for some disputes involving the patentability of a given subject matter, we must draw a distinction between an abstract idea and a concrete application of an idea. Roughly, the difference is that while an abstract idea is intangible and has no definite location in space and time, a concrete application is tangible and has some location in space and time. Thus, there is a difference between the idea of a telescope and a tangible telescope, because you can see with one but not with the other. Applications have definite uses. It takes some additional step of human ingenuity and effort to transform an idea into a tangible application. Philosophers, scientists, and science fiction writers had conceived of machines that could calculate and talk long before engineers and technologists developed the modern computer. Although the conception and design of an invention is one step closer to the invention than the abstract idea, it still falls short of the mark. A diagram of a computer is not a computer. Thus, a key issue in making the distinction between ideas and applications is deciding to what extent an abstract idea has been transformed into a practical application by human causes, such as ingenuity and effort.

The same issue, the role of human ingenuity, also plays a key role in making a distinction between the natural and the artificial. There is a sense in which all things in the world, including human beings and human inventions, are natural. Human beings are part of the natural world, and a house built by people is as natural as a robin's nest or a termite mound. A tool fashioned by man is just as natural as a tiger's tooth or a porcupine's quill. On the other hand, we can draw a distinction between natural things and artificial things (or artifacts) based on mankind's causal role in making, crafting, or using those

things. If human beings make, create, or fashion some thing, then it is an artifact. When we are talking about inventions, we can say that some type of
human intervention is required to transform natural processes and materials
into inventions. At the very minimum, an invention is a thing that would not
have existed without human causal agency.[3]

But there are many things that would not have existed in the world without human agency that we do not consider to be inventions. If I walk down
the beach and leave footprints in the sand, the footprints are the result of my
causal agency, but they are not inventions, even if we suppose that they have
some practical use and are not obvious. What type (or how much) of human
agency is required to transform a product of nature (or an abstract idea) into
a product of human ingenuity? Most people will agree on obvious cases: we
can all agree that X rays are a natural phenomenon but an x-ray machine is an
artifact; crude oil is natural but plastic is artificial. Fermentation is natural but
a method for making wine is not.

But what about more difficult cases, like DNA, genetically engineered
mice, or vitamin B_{12}? How can we draw a line between natural and artificial in
these cases? Are there any objective criteria for distinguishing between products
of nature and products of human ingenuity? It is my contention that there are
no such criteria and that we must appeal to normative concerns, such as our
goals, purposes, and values, in drawing this distinction. Consider a rock lying in
the woods. If I pick up the rock and move it, I have done something to it, yet
most us would not say that I have invented anything yet. But suppose I chip the
rock and make a statue or an arrowhead. Or suppose I take the rock and use it
as a doorstop. At what point does the rock become an artifact? What must I do
to the rock to transform it from a natural thing into a human invention?
According to the objectivist, there is some objective point at which the rock
becomes a human invention. If the item bears the correct causal relation to
human beings, the objectivist might argue, then it is an invention; if not, then it
not an invention. On this theory, we can sort all things into the categories natural or artificial, in the same way that we could sort all numbers into odd and
even, or sort all atoms into positive charge, negative charge, or neutral charge.

A moment's reflection can show us that the distinction between natural
and artificial is not objective nor is the distinction between product of nature
and product of human ingenuity. There is no definite quantity or quality of
human causal agency that is required to transform a rock from a product of
nature into a human invention (Resnik 2002). The rock is deemed a human
invention because, for certain purposes, we say that it should be called a
human invention. We could say that even a rock transformed into an arrowhead lacks sufficient human agency to call it an invention, or we could say that
a rock used as doorstop has sufficient human agency to make it an invention.
What we should say depends, for the most part, one our goals, purposes, and
values in calling the arrowhead or doorstop a human invention. At a bare min-

imum, an invention must be the result of human agency; but beyond that, we must decide what type of agency will suffice. Thus, the distinction between products of nature and products of human ingenuity is like other controversial distinctions in law and ethics, such as the distinctions between human and non-human, dead and alive, and competent, and incompetent (Resnik 2002a).

Further support for the conclusion in the preceding paragraph comes from some reflection on the pragmatic features of explanation (Garfinkle 1981; Risjord 2000). An explanation is an answer to a why-question, such as "Why did the car crash?" A successful answer will provide the person who makes the query with information that satisfies their request for explanation. Since people ask questions in order to satisfy different epistemological and moral goals and purposes, explanations must account for these normative aspects of language and thought. For example, if a car crashes on an icy road, we might consider several different types of causal factors in explaining the crash, such as the condition of the road, the condition of the tires, the possible negligence of the driver(s), and the presence of snow or ice. All of these different factors may indeed be causes of the crash in the sense that the crash would not have occurred without these factors. But which factor is the cause of the crash? A great deal depends on our purposes and goals in understanding and explaining the crash. If we are highway engineers, we may be interested in the condition of the road. If we are automobile insurers, we may be interested in the conduct of the driver(s). If we are tire manufacturers, we may be interested in the condition of the tires. The point I would like to draw from this example is that normative considerations play a key role in causal explanations and causal assertions. Since calling a thing an invention is a way of locating that thing within the casual nexus of the world based on its relation to human agency, we should also expect that normative considerations play a key role in the distinctions we make between human inventions and products of nature (Resnik 2002a).

Applying these points to the debate about DNA patenting, we can say that the question of whether DNA is a product of nature or a product of human ingenuity does not have an objective answer; the answer to this question is a function of how we decide to relate DNA to human agency. We could say that all DNA is natural, or we could say that some types of DNA, such as isolated and purified DNA, are products of human ingenuity. As a bare minimum, some type of human agency is required in order for us to call DNA a human invention. So, we should not call naturally occurring DNA a human invention because human agency is not implicated in the production of naturally occurring DNA. But once human beings take an inventive step to isolate DNA from nature or modify natural DNA, we have sufficient causal agency to say that this type of DNA is a human invention, if we so choose. Once the bare element of human causal agency is established, we must decide what types of human-produced DNA should be classified as patentable, human inventions. Where we decide to draw this line does not depend on objective features of the world but on how we decide to think about the place of DNA in the world.[4] Thus, to decide whether DNA is a product of human ingenuity, we need to ask ourselves how our classification of DNA relates to our goals, purposes, and values. If we take a utilitarian approach to this inquiry, we should evaluate the classification scheme based on its consequences.[5] Thus, in deciding whether isolated and purified DNA should be viewed as a product of nature or a product of human ingenuity, we should ask ourselves how calling isolated and purified DNA a human invention will affect science, medicine, agriculture, economic growth, respect for human life, and so on. As we saw earlier, this utilitarian perspective can help us to understand why it is wrong to patent nature and why it is not wrong to patent human inventions. I happen to accept the utilitarian approach as a useful strategy for formulating and defending public policy, but I will not defend it in depth here (see Scheffler 1988). I hope to show in this book that the likely good consequences of DNA patenting justify the risks they pose. In weighing and assessing the consequences of our classification schemes concerning DNA (and other biological materials) we must strike a balance between public access and ownership private dominion and control. Currently, the U.S. legal system has struck a balance that places naturally occurring DNA sequences in the public domain but allows private control of isolated and purified (or modified) DNA sequences. In this book, I hope to show that this is basically a good way of balancing public and private control, even though it may need occasional tweaking to maintain the correct balance.

Even if one does not take a utilitarian approach to ethics and public policy, the choice of how to classify DNA depends, at least in part, on the consequences of calling DNA a product of nature or a product of human ingenuity, since any moral theory worth taking seriously gives some consideration to consequences even if it is not regarded as a consequentialist theory (Rawls 1971). For these other approaches, the main question would be whether there are some deontological constraints on classifying some types of DNA as human inventions. Earlier in this chapter, I examined and critiqued several deontological arguments against classifying some forms of DNA as human inventions. In the next chapter, I will examine and critique another deontological argument against classifying some types of DNA as human inventions. As a preview to these arguments, I hope to show that there are no deontological constraints on classifying isolated and purified (or modified) DNA sequences as products of human ingenuity, although there are some constraints on declaring a whole human genome to be a human invention.

CONCLUDING REMARKS

This chapter has defended the position taken by the PTO and the EPC that we can consider some types of DNA as patentable, products of human of ingenuity, although we should not consider DNA in its natural state to be patentable. Before concluding this chapter, I would like to consider important objections to this view.

Objection: How can you draw a distinction between natural DNA and invented DNA when natural DNA and invented DNA are structurally and functionally identical? Moreover, since invented DNA is nothing more than a copy of natural DNA, patents on invented DNA will give people control over natural DNA. Thus, DNA patents really are an attempt to own nature.

Reply: Invented DNA is not structurally and functionally identical to natural DNA. During the process of isolating and purifying natural DNA, organic processes change the DNA so that it is not identical to its naturally occurring form. Scientists also introduce specific changes, such as removing or adding DNA sequences. To make complementary DNA (cDNA), one removes sequences that do not code for proteins. Scientists may add nucleotide sequences to DNA in order to make modifications in protein products. Even when scientists make no changes to isolated and purified DNA, DNA in its purified form is not identical to DNA as it exists in nature because in nature it always exists in an impure form. If it makes sense to say that isolated and purified vitamin B_{12} is not identical to naturally occurring vitamin B_{12}, then it also makes sense to say that isolated and purified DNA is not the same as naturally occurring DNA.

On the other hand, this objection raises an important point that we should not forget: if we want to say that isolated and purified DNA is a patentable human invention but that naturally occurring DNA is not patentable, then we must ensure that DNA patent holders are not allowed to enforce any claims on naturally occurring DNA. A person who studies the naturally occurring form of a patented DNA sequence should not be in fear of patent infringement for using the naturally occurring version of the DNA. Thus, these concerns can be framed as a question about the scope of a DNA patent: a DNA patent should be interpreted as granting the inventor exclusive rights over artificial DNA but not rights over natural DNA.

For example, Amgen has a patent on isolated and purified DNA sequences that encode erythropoietin. Someone else who uses, makes, or sells this isolated and purified DNA without Amgen's permission can be sued for patent infringement. However, someone who uses, makes, or sells organisms or cell lines that contain this DNA sequence should be protected from a patent infringement claim on the grounds they are not using, making, or selling the gene in its purified form. The scope of Amgen's patent should not extend to natural DNA. The laws should allow researchers to continue to study natural DNA without the fear of legal repercussions.

Objection: Aren't you assuming a reductionistic view of life? Doesn't a reductionistic view of life threaten the value we place on life, especially the value we place on human life?

Reply: It depends on what one means by reductionism. Reductionism is roughly the view that complex phenomena at a higher level of organization can be explained by phenomena at a lower level of organization. For example, a reductionist approach to chemistry holds that chemical reactions can be explained by interactions that occur at the sub-atomic level of organization. Hydrogen and oxygen combine in a 2 to 1 ratio because each hydrogen donates one election and each oxygen accepts two electrons. A reductionist approach to biology attempts to explain the behavior of complex biological systems, such as organisms, in terms of their simpler parts, such as organs, cells, organelles, membranes, proteins, DNA, and so forth. In biology, there a many different hierarchies of organization from, ranging from the molecular to the cellular level, on up to the anatomical and physiological levels (Rosenberg 1985).

Prior to the twentieth century, a view known as vitalism had a great deal of influence on biologists. According to vitalists, complex organic phenomena, such as metabolism, reproduction, and development, cannot be explained in terms of the more basic parts of organisms, because all living things contain a vital force or *élans vital*. Although most biologists no longer accept vitalism, many biologists, known as organicists, hold that living things have emergent properties that cannot be explained by referring only to their component parts. Emergent properties are properties that can only be explained by reference to the whole organism and its properties and parts (Mayr 1982). For example, one might argue that complex behaviors, such as mating, migration, and foraging are emergent properties because they cannot be explained in terms of the component parts of the organisms (i.e., cells, genes, and proteins). Organicists do not believe in any vital forces or entities. They hold that organisms are composed of various chemicals, including DNA, and that many organic processes can be explained by chemical processes. However, they deny that one can understand or explain all biological phenomena by referring only to chemical phenomena.

In genetics and genomics, reductionism is associated with genetic determinism (i.e., the idea that complex biological phenotypes can be explained by genetic causes). Although genetic determinism has become increasingly popular among laypeople in the last three decades, most biologists reject genetic determinism. All phenotypes result from genetic, environmental, and developmental causes. For example, cancer is caused by a number of different factors, including radiation, exposure to chemicals, viruses, and genetic predispositions. A person could have a genetic predisposition for cancer and yet not develop cancer. Likewise, a person could develop cancer without having any significant genetic predisposition for the disease.

How do these issues relate to DNA patenting? Those who maintain that DNA can be patented hold that DNA is a type of chemical; it is not a mysterious vital entity or some irreducible part of a greater whole. As a chemical,

DNA interacts with other chemicals in the cell, such as RNA and proteins. These interactions can affect the behaviors of cells, which interact with other cells and tissues to influence physiology and behavior. Accepting the patenting of DNA does not commit one to a reductionistic approach to life, since one could hold that DNA is a chemical yet also maintain that not all biological phenomena can be explained in terms of the interactions of chemicals. DNA patenting is perfectly compatible with organicism. DNA patenting also does not imply genetic determinism, since one can hold that DNA is a chemical that plays an important role in causing phenotypes but also hold that other causal factors, such as the environment, play an important role in causing phenotypes. DNA patenting need not imply biological reductionism or determinism (Peters 1997).

Since DNA patenting does not entail reductionism, it does not, by itself, threaten the value we place on life. So how could reductionism affect the value we place on life? A reductionistic approach to the value of life would hold that the value of a living thing is no more than the sum of the value of its component parts. For example, one might hold that the value of a cat is equal to the value of its proteins, DNA, calcium, lipids, and other parts. Few people accept reductionistic approaches to value because people make value judgments holistically. For example, the value of a book is more than the value of its individual pages; the value of a symphony is more than the value of its notes, and the value of a house is more than the value of its boards, nails, and bricks. Indeed, there are very few things that we do not value holistically. We probably value money reductionistically; the value of the money in my pocket is simply the sum of the values of all of the coins and dollar bills in my pocket.

As one can see, DNA patenting does not imply any of these reductionistic approaches to the value of life. Not only does DNA patenting not imply biological reductionism, it also does not imply axiological reductionism. One can accept the idea that DNA is a chemical without accepting the idea that an organism (or a human being) is worth nothing more than the value of his or her chemicals. One can accept the idea that DNA has some commercial value without holding that the sole value of a human being resides in the value of his or her patented parts. Granted, DNA patenting may encourage reductionistic attitudes toward the value of life, and it may even encourage cultural infatuation with genetic determinism, but it does not entail either of these results. DNA patenting is perfectly consistent with the idea that living things have irreducible moral or social value. We will return to these issues again in chapter 6.

6

DNA Patents and Human Dignity

INTRODUCTION

This chapter will examine and critique an important deontological challenge to DNA patenting, namely, the objection that DNA patents violate human dignity. We have already seen how members of the "Joint Appeal against Human and Animal Patenting" compared DNA patenting to slavery, while many critics have objected to treating human beings (or their parts) as marketable commodities (Rifkin 1998; Andrews and Nelkin 2000). Kimbrell (1997) launches a particularly vociferous condemnation of the commodification of all biological materials, including DNA. I will let his words speak for themselves:

> The patenting of human genes by government and private corporations could create a unique and profoundly disturbing scenario. The entire human genome, the tens of thousands of genes that are our intimate common heritage, would be owned by a handful of companies and governments. (Kimbrell 1997, 228)

> We are now in the early stages of adding the human body, its parts and processes, to the list of commodities that are subject to the laws of supply, demand, and price. The body is not a commodity. It is not a manufactured product intended for consumption. . . . The new techniques in biotechnology . . . are now leading to the commodification of the body. . . . Animals are being genetically mutated, often cruelly, in order to make them better products or factories for the

93

production of valuable human genes and chemicals. And corporations are poised to gain patent ownership of all human genes and
valuable cells. (Kimbrell 1997, 330–31)

Although Andrews and Nelkin are more circumspect than Kimbrell in their
analysis of DNA patenting, they also are concerned about the threat to human
dignity posed by patenting:

> In the current genetic gold rush, patients and the public are losers.
> Costs of genetic diagnostic tests have skyrocketed and research has
> slowed. But more tragic is the shift in the cultural value of people. No
> longer valued as active producers of the product, they have become
> the raw material itself. . . . That human beings would be perceived as
> raw materials was predicted twenty years ago in a brief filed in the
> *Chakrabarty* case. (Andrews and Nelkin 2001, 63)

The European Commission also weighed in on this issue when it stated
that the EPC may refuse to patent an invention that infringes on the rights of
the person or violates human dignity (European Commission 1998). As noted
in chapter 3, the EPC has a public morality clause that allows the agency to
refuse patents that are against the public morality. There has been considerable discussion among members of the EPC about what types of biotechnology patents would be against the public morality (Crespi 2000). Directive
98/44 of the European Parliament and Council declares that patents on
processes for cloning human beings, processes for modifying the germ-line of
human beings, and uses of human embryos for commercial purposes would be
against the public morality (European Parliament and Council 1998). The
main moral objection to these types of patents is that they would violate
human dignity by treating people as manufactured goods or commercial products (Knoppers, Hirtle, and Glass 1999).

In chapter 3 I also noted that the Council for Responsible Genetics
opposes DNA patenting in its Genetic Bill of Rights. In the preamble of this
document, the Council states that manipulation of human genes threatens
human rights and dignity (Council for Responsible Genetics 2000). During
his unsuccessful presidential campaign, Green Party candidate Ralph Nader
used gene patenting as yet another example of how corporations have run
amok and have commercialized things that should not be for sale, including
human beings (Nader 2000).

Thus, many different critics, scholars, consumer advocates, religious leaders, and some scientists have argued that DNA patents violate human dignity.
On the other hand, scientists, patent attorneys, and business leaders have
found it hard to understand why people would think that DNA patents violate human dignity (Evans 1999). If DNA is a chemical compound, then how

could DNA patents pose a threat to human dignity? A patent on DNA should pose no more of a threat to human dignity than a patent on isolated and purified insulin, growth hormone, or genetically engineered clotting factors. However, since DNA encodes genetic information, it is no ordinary macromolecule. As we noted in the previous chapter, some people regard the human genome as our common heritage. As we shall see in this chapter, some people regard the human genome as intimately related to personal identity.

If one could show that some types of DNA patents violate human dignity, then this would constitute a strong deontological argument against those types of patents, since it is wrong to violate human dignity. In this chapter, I shall argue that only a patent on a whole human genome would actually violate human dignity, since this type of patent would treat a whole human being as complete commodity. Other DNA patents, which would treat human beings as incomplete commodities, would threaten but not violate human dignity.

HUMAN DIGNITY AND MARKET RHETORIC

In order to understand how DNA patents might violate or threaten human dignity, it is important to think about the moral basis for human dignity. We need to distinguish between intrinsic or inherent values and extrinsic or instrumental values (Frankena 1973). Intrinsically valuable things are valuable for their sake; extrinsically valuable things are valuable for some other end or goal.[1] For example, many people hold that happiness is intrinsically valuable, although a dental appointment is only extrinsically valuable. It is possible for something to be both intrinsically and extrinsically valuable. For example, we might value health for its own sake and also as a means to other goals, such as wealth or happiness.

The claim that human beings have moral dignity is an assertion about how we should treat people: we should treat people as valuable for their own sake, not as mere means to other ends or goals. Many different theologians and moral philosophers have defended the idea that people are intrinsically valuable. The eighteenth century German philosopher Immanuel Kant (1724–1804) held that human beings have intrinsic moral dignity or worth because they can choose to make their conduct conform to the foundational moral principle known as the Categorical Imperative (CI). Kant held that one can derive all moral rules and imperatives from the CI. According to a version of the CI known as the formula of the end in itself, we have a moral duty to treat all human being as ends in themselves not merely as a means to other ends:

> Persons are, therefore, not merely subjective ends, whose existence as
> an effect of our actions has a value for us; but such beings are objec-
> tive ends, i.e., exist as ends in themselves. . . . The practical imperative

will therefore be the following: Act in such a way that you treat
humanity, whether in your own person or in the person of another,
always at the same time as an end and never simply as a means. (Kant
1981, 36, sec. 429)

Kant illustrates his principle with two examples, suicide and making a
false promise. Suicide is immoral because killing one's self in order to escape
a difficult situation treats one's own person as simply a means to another end.
Making a false promise, a promise that one does not intend to keep, is
immoral because it treats another person as simply a means to another end
(Kant 1981, 36–37, sec. 429).

Kant argues that we should value people for their own sake (i.e. as objective ends), not for the sake of some other ends (i.e., as subjective ends). We
should not treat people as objects or things. The reason we should act this way,
according to Kant, is that all rational beings necessarily regard themselves as
intrinsically valuable and therefore should also regard other rational beings as
intrinsically valuable (Kant [1785] 1981, 36, sec. 429). Because rational beings
have free will, they generalize from the intrinsic worth of their own existence
to the intrinsic worth of other people. There are two types of freedom, according to Kant, negative freedom and positive freedom. Negative freedom consists in freedom from external control, while positive freedom consists in
bringing one's will in conformity with moral laws that are valid for all rational beings:

What else, then, can freedom of the will be but autonomy, i.e. the
property that the will has of being a law unto itself? . . . A free will
and a will subject to moral laws are one and the same. (Kant [1785]
1981, 49, sec. 446)

Thus, on Kant's view, a rational being exercises his or her positive freedom
when he or she treats all people as intrinsically valuable. He or she regards her
own moral worth as holding for all people.

Contemporary Kantians (Hill 1992; Korsgaard 1996) have endorsed versions of the CI, and have elaborated on the concept of moral agency. Moral
agents (or persons) are beings capable of understanding, developing, and following moral rules. Moral agents have moral duties and moral rights with
respect to other moral agents in the moral community. Individuals that are
developing into moral agents, such as children, or those who cannot become
moral agents, such as severely mentally retarded people, may still have moral
rights because they have interests, which members of the moral community
regard as worth protecting. Kantians also hold that moral worth is held
equally: one person does not have greater moral value than another person. If
moral worth were distributed unequally among people, then it would be eth-

ical to kill an innocent person in order to maximize overall moral value, a proposition that most Kantians would reject.[2]

Various religious traditions, including Christianity, Judaism, Islam, and African and Native American religions, also provide support for the idea that human beings are intrinsically valuable (Ganeri 1997; Peters 1997). Since I cannot give a reasonable account of all the world's religions here, I will focus on what the Bible has to say about human dignity, since three of the world's major religions, Judaism, Christianity, Islam accept some of the teachings and precepts found in the Bible. (Jews and Christians both accept the Old Testament, but Jews do not accept the New Testament. Muslims believe in the prophets of The Bible but they also believe in the prophet Muhammad's teaching in the Quran.) In the very beginning of the Bible, one finds an account of human beings as created in the image of God:

> And God said, let us make man in our image, after our likeness: and let them have dominion over the fish of the sea, and over the fowl of the air, and over the cattle, and over all the earth, and over every creeping thing that creepeth upon the earth. So God created man in his *own* image, in the image of God created them, male and female God created them. . . . And God saw everything he had made, and, behold, it was very good. (Gen. 26–27, 31)

According to many Biblical scholars, the claim that man is created in the image of God does not mean that the human body is made in God's image, but that God endowed human beings with moral, spiritual, emotional, and cognitive attributes that reflect His divine nature. Our moral worth derives from our relationship with God. Since God is intrinsically good, and human beings are made in His image, human beings are intrinsically good. Human beings are also equally worthy because all people are made in the image of God (Brunner 1947).

Other influential passages in the Bible's Old Testament speak of the relationship between God and human beings. Psalm 23 compares the relationship between God and His people to the relationship between a shepherd and his sheep:

> The Lord is my shepherd; I shall not want. He maketh me to lie down in green pastures: he leadeth me beside the still waters. He restoreth my soul; he leadeth me in the paths of righteousness for his name's sake. Yea, though I walk through the valley of the shadow of death, I will fear no evil: for thou art with me; thy rod and thy staff they comfort me. Thou preparest a table before me in the presence of mine enemies: thou anointest my head with oil; my cup runneth over. Surely goodness and mercy shall follow me all the days of my life: and I will dwell in the house of the Lord for ever. (Ps. 23:1–6)

Other passages in Psalms describe the relationship between God and His people, and other books in the Bible, such as Exodus, Deuteronomy, Joshua, II Samuel, Job, Jeremiah, and Daniel also speak of a God who cares for His people.

Jesus' teachings in the Bible's New Testament elaborate on the Old Testament's description of the relationship between God and His people. Whereas the Old Testament emphasized God's justice and righteousness, the New Testament emphasizes God's mercy and compassion. Alluding to Psalm 23, Jesus compares himself to a shepherd:

> I am the good shepherd, and know my sheep, and am known of mine. As the Father knoweth me, even so knoweth the Father: and I lay down my life for sheep. And other sheep I have, which are not of this fold: them also I must bring, and they shall hear my voice; and there shall be one fold and one shepherd. (John 10:14–15)

Here, Jesus also emphasizes the value and worth of human beings (sheep), since people are so valuable that the shepherd (God) will sacrifice his life for his sheep.

Jesus' teachings about love can also be interpreted as an injunction to treat God, one's self, and other people as having intrinsic worth:

> Then one of them, which was a lawyer, asked him a question, tempting him, and saying, Master, which is the greatest commandment in the law. Jesus said unto him, Thou shalt love the lord thy God with all thy heart and all thy soul and all thy mind. This is the first great commandment. And the second is like unto it, thou shalt love thy neighbor as thy self. (Matt. 22:37–39)

To love, in this Christian sense, is to love unconditionally, and to love something unconditionally is to treat it as valuable for its own sake (Brunner 1947).

The story of the prodigal son (Luke 15:11–32) provides a good illustration of this unconditional love. In this story, a man has two sons. The older son stayed home and kept his father's commands. The younger son left home and asked his father for his inheritance. While away from home in a foreign country, the younger son "wasted his substance with riotous living" (Luke 15:13). The prodigal son soon discovered the error of his ways and returned home, seeking forgiveness. Instead of holding a grudge against his son, the father "had compassion, ran, and fell on his neck and kissed him" (Luke 15:20). In this story, the father loves his son and regards him as deserving of love, even though the son strayed from the fold.

It is no accident that Kant's moral philosophy has so much in common with the Biblical tradition, since Kant belonged to the Lutheran church and sought to provide a philosophical justification for Christian moral imperatives, such as the Ten Commandments, the Golden Rule, and Jesus' teaching about love (Green 2001).

Thus, Kantian philosophy and the Bible both support the general idea that people have moral dignity and worth and should be treated as valuable for their own sake. However, philosophers and theologians do not agree about what it means to treat people as valuable for their own sake or what it might mean to violate human dignity (O'Neill 1996). Does committing suicide in order to save other people violate one's own dignity? If so, is it immoral? Many people would accept the idea that a person who jumps on a hand grenade in order to save five other people has not violated the moral law and has committed a praiseworthy act. Although the Judeo-Christian tradition also condemns suicide, few Christians would condemn suicide to save other people, since the good shepherd sacrifices his life for his sheep.

What about amputating one's own leg in order to save one's life? If a person has severe gangrene in his leg, it may be necessary to amputate his leg to save his life. Does amputation treat one's own person as a mere means to one's own life? Clearly, neither Kantians nor Christians want to hold that amputation is morally wrong. Indeed, Matthew (5:29–30) instructs believers to cut out their right eye or right hand if it is offensive, since it is better to lose a part than the whole body. What about mutilating or changing one's own body for the sake of cultural and aesthetic value (e.g., body piercing or tattooing). Do these actions violate human dignity? For thousands of years the Jews have accepted the practice of circumcision as a sign of one's devotion to God. Christians have also accepted self-mutilation as symbolic of religious devotion. One way of making sense of these practices that damage or destroy body parts is to say that they can be justified to serve a worthwhile goal, such as saving one's life or one's soul.

To make sense of the idea that destruction of a body part might not violate human dignity, we need to say a bit more about the concept of personhood and the idea of treating something as intrinsically valuable. Is a person simply a rational will? Is a person an embodied will? What is the relationship between the person and his or her body? Does treating a person as extrinsically valuable preclude treating that person as intrinsically valuable? Can you treat someone as both extrinsically and intrinsically valuable at the same time? These are important questions for moral philosophy to address, quite apart from their relevance to Kantian philosophy. I will return to these questions shortly.

Before returning to these issues, we need to explain the connection between human dignity and market rhetoric. How could selling a body part (as opposed to destroying it) violate human dignity? Let us begin by distinguishing between market values and nonmarket values. A market value is an economic measure of value, such as a price. To put a price on a thing is to attribute a property to the thing that is a measure of its market worth. If we accept the premise that money is not intrinsically valuable but only extrinsically valuable, then market values are extrinsic values. Price is not a measure of the intrinsic worth of a thing but is a measure of its extrinsic worth. To assign a price to a thing is equivalent to treating that thing as a type of commodity. Now we can see the connection between human dignity and market rhetoric. Treating a person as intrinsically valuable requires us to treat that person as having a value that cannot be measured in markets terms (Radin 1996). If we treat a person as having only a market value, then we have treated that person as a commodity.

But can we treat something as having both a market value and a nonmarket value? Or, to reframe the question, can we treat something as both intrinsically and extrinsically valuable? Recall that Kant does not say that one can never treat a person as a mere means; one may treat persons as a mere means and as an ends in themselves at the same time. This premise implies that it is morally acceptable to treat a person as having a market value, provided that we also treat him as having a nonmarket (or moral) value. Consider a priceless painting that we assign a value of $10 million for insurance purposes, based on the price it is likely to fetch on the market. Although we have treated the painting as having a market value, we also realize that this price is not a measure of its true worth (i.e., its aesthetic or cultural value). Likewise, one could insure a human being for $1 million yet also hold that the insurance policy is not a measure of that person's true worth (i.e., his or her moral worth).

As we saw in chapter 3, to commodify (or commercialize) something is to treat it as an object having a market value. Recalling the distinction between complete commodities, incomplete commodities, and noncommodities, we can now say that complete commodities are treated as having only a market value, incomplete commodities have a market value as well as some other value, and noncommodities are treated as not having a market value. Noncommodities are taken off the market entirely and are not for sale. Complete commodities can be bought, sold, manufactured, standardized, manipulated, and so forth.

HUMAN DIGNITY AND DNA PATENTING

We can now frame the objections to DNA patenting in relation to human dignity: those who argue that DNA patents violate human dignity claim that these patents treat human beings as complete commodities, as mere things with commercial value but with no intrinsic moral worth (Resnik 2001b). Human beings should not be treated as having only a market value; they should not be treated as complete commodities (Walzer 1983). Ergo, the patenting of human DNA is immoral.

Before we consider this type of objection to DNA patents, it is important to note that we currently live in a society that treats human beings as incomplete commodities, not as pure, noncommodities (Resnik 2001b). Although we view people as having intrinsic moral worth, we also apply market rhetoric to human beings in many ways. Consider the following practices that put a price on human life:

- Insurance policies that provide compensation for the loss of life, loss of limb, or disability;
- Wrongful death lawsuits, which assign an economic value to a person's life;
- Negligence lawsuits that compensate victims for injury or pain or suffering;
- Cost-benefit analyses conducted by regulatory agencies attempt to balance lives lost or saved against the economic costs of improving the safety of medicines or new food products.
- Wages, salary, and other forms of compensation for human labor, which put a price on human activity.
- Patents, copyrights, and trademarks, which put a price on human ideas.

Indeed, it is hard to find any area of human life in western, capitalistic societies that is completely immune from the invasion of market rhetoric and market values. As long as we have private property, democracy, and a free market economy, it is difficult to see how we could ever shield human beings from the market. Indeed, one of the key pillars of capitalism is the idea that free market economies are more effective at distributing goods and services and stimulating economic growth and innovation than centrally planned economies (Samuelson 1980).

However, one might object to this apathetic attitude toward commodification in society and even though currently society partially commodifies human beings, we should not abide by this practice. Society's traditions and practices could be morally corrupt. Instead of passively accepting these practices, we should rebel against them and reform society. Indeed, according to some writers, Kant condemned all marketing of the human body, including prostitution and even the selling of one's teeth, because all marketing of one's body treats one's own person as a mere means (Morelli 1999; Green 2001). Kant thought the body should be treated as a noncommodity. Others have argued that Kant only condemned the commercialization of essential body parts (Cohen 1999). Still others have argued that Kant would accept the selling of kidneys because "my kidney is not my humanity" (Gill and Sade 2002, 26).

For the rest of this discussion I will focus on Kant's attitude toward the body rather than the biblical tradition, since I do not consider myself to be Biblical scholar, although I do claim some expertise in Kantian philosophy. I will attempt to answer some questions about Kant's attitude towards commodification of the body. As noted above, answering these questions requires one to explicate Kant's understanding of the connection between the person and his or her body and to specify what it means to treat a person as a means merely. Apparently, Kant did not oppose the amputation of one's limbs to save one's own life, since right after discussing suicide he says in a parenthetical remark, "A more exact determination of the principle so as to avoid all mis-understanding (e.g., regarding the amputation of limbs in order to save one-self)" (Kant [1785] 1981, 36, sec. 429). This remark suggests that it is some-times acceptable to treat a part of one's own body as a mere means, since amputation sacrifices a part of the body for the sake of the whole body. But how can Kant say that it is unacceptable to sell one's own teeth but that it may be acceptable to amputate a limb?

Perhaps the best way to resolve this apparent inconsistency in Kant's phi-losophy is to apply a different version of the CI to these two propositions. According to the universal law formula of the CI, one should "Act as if the maxim of your action were to become through your will a universal law of nature" (Kant [1785] 1981, 30, sec. 421). To apply this principle to a particu-lar maxim (i.e., a rule for action), we must ask ourselves first, Could the maxim become a universal law?, and second, Would a rational being will it? If we con-sider the maxim, I will amputate a leg in order to save my life, we can see that this maxim passes the universal law test because this could become a univer-sal law, and a rational being would will this maxim, since a rational being would recognize that he or she might sometimes need to save his or her life. The other maxim, I will sell my teeth for money, also passes the first part of the test, since it is certainly possible that everyone could follow such a rule. But would a rational being will this maxim? Perhaps that answer to this ques-tion depends on how a rational being would regard his or her teeth. Clearly, a rational being would not sell his or her heart (or some other vital organ) for money, since this action would sacrifice his or her life for money. A rational being would also not sell him or herself in slavery. Both of these actions would be immoral because they would treat one's own person as a valuable only for the sake of some other end. But what about selling of teeth and other nonessential body parts, such as blood, sperm, and hair? I am not sure how Kant would answer this question. One wonders if some of Kant's remarks about the selling of teeth and sexual services stem from his moral principles or merely reflect prejudices based on the Lutheran tradition that Kant attempted to rationalize (Green 2001). According to Green:

> Like most educated Christians of the eighteenth century, Kant believed that suicide, extramarital sex, masturbation, and homosexual acts are morally wrong. Indeed, he termed the latter "unmentionable vices" and held that they represent a violation of the humanity in our own person ... Kant may have been a radical innovator in moral the-ory and a pioneer when it came to understanding the rational struc-ture of moral reasoning. But in practical ethics, he was a profound conservative. (2001, 252)

This quote from Green underscores some of the difficult historical problems related to determining Kant's exact position on issues relating to the commodification of the body.

One final way of approaching the problem would be to appeal to Kant's distinction between perfect and imperfect duties (Hill 1992). A perfect duty embodies an absolute moral rule that one should never violate. Not committing suicide is a perfect duty to one's self; not making a false promise is a perfect duty to others (Kant [1785] 1981, 30–31, sec. 422). Imperfect duties, on the other hand, are conditional; one may violate an imperfect duty in order to fulfill a different duty. Cultivating one's talents is an imperfect duty to one's self, since one may set aside this duty in order to help someone else. Likewise, helping a particular person is an imperfect duty to another person, since one may neglect this duty in order to help one's self or another person (Kant [1785] 1981, 30–31, sec. 422). Since imperfect duties may conflict, one must use moral judgment and discretion to decide whether, how, and when to honor an imperfect duty. If the selling of an essential body part is analogous to slavery or suicide, then it follows that one has a perfect duty not to sell an essential body part. If selling a nonessential body part is like failing to cultivate one's talents, then one only has an imperfect duty not to sell a nonessential body part. Other things being equal, one should not sell nonessential body parts, but the selling of nonessential body parts could be justified in order to fulfill some other duty, such as the duty to save a human life or perhaps even the duty to promote the progress of science.

I think this distinction between perfect and imperfect duties does help us make some headway in developing a Kantian approach to the problem of commodifying the human body. However, it still does not help us solve the problem, with the exception of essential body parts, of whether there are some parts that should not be for sale. Where can we draw the line between acceptable and unacceptable commodification of the body? In the final analysis, I think the best we can say is that Kant regarded complete commodification of the body as immoral and that he some moral qualms about various forms of incomplete commodification.

If we move beyond Kant's view and consider a more libertarian approach, we can find ample justification for the commodification of the body. John Locke, the founder of libertarian political philosophy, held that our bodies are our own property:

> Though the earth, and all inferior creatures, be common to all men, yet every man has *property* in his own *person:* this no body has right to but himself. The *labour* of his body and the *work* of his hands, we may say, are properly his. (Locke [1764] 1980, 19, chap. 5, sec. 27)

As noted in chapter 3, although Locke accepted the idea of common property given to all, he also believed in private property. One way to acquire private property is to mix one's labor with common property given to all. This libertarian emphasis on property and property rights helped to provide a philosophical justification for capitalism, private property, free markets, and the U.S. Constitution. The great economist Adam Smith (1727–1790) followed in this libertarian tradition and laid the foundations for capitalism. In his *Wealth of Nations* ([1776] 1991), Smith articulated the concepts of supply, demand, price, the division of labor, wealth, trade, and markets.

One of this book's working assumptions is that some form of regulated capitalism will operate in society for the foreseeable future and is morally justified. Although some extreme libertarians endorse unfettered capitalism, the more sensible view is that capitalism needs some form of government regulation. Regulation is important to prevent activities that hinder and corrupt the free market, and also to promote important social values, such as justice, public health and safety, and environmental quality and integrity. Regulated capitalism has proven itself economically and morally superior to other ways of organizing human production and labor, such as socialism, feudalism, tribalism, or laissez faire capitalism.

Given these assumptions, I can conceive of no way of treating people as pure, noncommodities in a capitalistic economy. Under capitalism, agents are free to exchange their labor for commodities, and they are free to exchange commodities for other commodities. The true value of a commodity is the amount of labor or other commodities it can be exchanged for, but it is more convenient to exchange commodities or labor for money. Although a capitalistic society could exist without money, it must have some way of setting a market value for those things that are exchanged (Smith [1776] 1991). If a person exchanges her labor for a commodity or for money, then that person is incompletely commodified. According to western legal systems, many types of rights, such as property rights or contractual rights, can also be exchanged for commodities (Calamari and Perillo 1998). Unless we decide to do away with capitalism and adopt some other type of economic system, we should expect that people will treat each other as incomplete commodities.[3]

Given that we already treat people as incomplete commodities, one can understand why some critics of practices that commodify people would be concerned about the risks of additional commodification of human beings. As we saw in chapter 3, incomplete commodification raises the specter of a slippery slope toward complete commodification: once we place an item on the market, it is difficult to stop or control the corrupting influence of money and commercial interests (Radin 1996; Hanson 1999). Examples abound of various social practices that have slid down the slope from incomplete commodification toward complete commodification, such as politics, sports, human reproduction, and the law.[4] For example, consider the increasing role of money in sports. At one time, most athletes were amateurs and participated in sports in order to develop moral virtues, such as courage, justice, perseverance, fairness, sportsmanship, self-discipline, and integrity. Indeed, the ancient Greeks regarded athletics as of supreme importance in moral development. Attitudes toward participation in athletics have changed in the last century as sports have become increasingly commercialized. Money now plays a crucial role in major sports at various levels of competition. Even high school athletes, especially basketball players, can be influenced by the prospect of financial gain. As money has continued to invade the world of athletics, people have become concerned that its influence will threaten or destroy important moral values essential to sports, such as fairness, integrity, honesty, and sportsmanship. Although most sports retain many of their nonmarket values, some sports, such as professional boxing or horse racing, have become highly commercialized. In order to stop the slide toward complete commodification, many sports have instituted rules governing financial relationships, contracts, endorsements, and gambling. The implications for DNA patenting from this example are quite clear: if DNA patenting constitutes incomplete commodification of human beings, does it create a distinct risk of additional commodification? Would additional commodification threaten human dignity? Does DNA patenting constitute complete commodification of human beings? Would complete commodification violate human dignity? We will now turn to these questions.

Violations versus Threats

In considering the human dignity argument against DNA patenting, I would like to distinguish between two distinct claims about the patenting of DNA:

- DNA patenting *violates* human dignity because it constitutes complete commodification of human beings.
- DNA patenting *threatens* human dignity because it could contribute to additional commodification of human beings and push us down a slippery slope toward complete commodification of people as well as other abuses associated with commodification, such as exploitation.

Violations and threats are morally distinct actions. I can threaten a person's rights to free speech without violating that person's rights. I can threaten the balance of an ecosystem without violating that balance. Likewise, I can threaten a person's dignity without actually violating that person's dignity. For example, raping a person violates that person's dignity because it treats that person as only a means to some other end, such as sexual gratification, domination, or revenge. A movie that glorifies rape threatens but does not necessarily violate human dignity. The movie could threaten human dignity by encouraging or enticing people to commit rape. Selling someone as a slave violates that person's dignity. Exploitative labor practices, such as child labor, threaten human dignity because they could lead to practices that are equivalent to slavery.

Most people in society regard actions that violate human dignity, such as murder, rape, assault, battery, theft, robbery, fraud, and slavery as immoral. Most of these actions are also illegal. But should we consider an act to be immoral or illegal because it merely threatens human dignity? The list of activities that threaten, but do not violate, human dignity includes pornography, violence on television and movies, advertising, fraternities and sororities, human experimentation, professional athletics, gambling, drug and alcohol use, religion, free speech, sexual relationships, and paid labor. Since many of these activities have some redeeming social value and they involve personal liberties, societies that value personal freedom, such as the United States, have good reasons for allowing people to engage in these practices. Quite often, a country will respond to threats to human dignity by formulating laws or policies that are designed to protect human rights and welfare without imposing undue burdens on personal liberty. For example, U.S. pornography laws reflect some balancing of liberty interests and social values. Pornography poses a threat to human dignity, but the United States has decided to adopt laws that attempt to minimize this threat while preserving liberty interests, such as the right to free speech.

Many of the laws designed to minimize threats to human dignity serve to prevent abuses and counter slippery slope concerns. For example, many communities have laws and standards that regulate the sale of pornography. These laws are designed to protect the community's values and prevent exploitation. Laws concerning campaign financing and gifts to politicians are designed to protect the political system from the corrupting influence of money. (Whether these laws actually work is a separate question, of course.) Laws banning the sale of human organs can also be viewed as restrictions designed to counter potential threats to human dignity posed by organ selling, such as economic exploitation (Andrews and Nelkin 2001; Gill and Sade 2002).

How should we decide how to respond to a purported threat to human dignity? There are three basic approaches:[5]

1. Ban or forbid the activity. For example, organ selling is illegal in the United States. Certain forms of gambling are illegal in most states in the United States.

2. Allow the activity to occur with little or no regulation. For example, there is very little regulation of violence on television in the United States, despite studies linking the viewing of violence on television to criminal activities.

3. Regulate and control the activity. As noted above, the United States has regulations on pornography, professional athletics, gifts to politicians, and so on.

REASONABLE VERSUS UNREASONABLE RISKS

In order to decide which option to take, we must consider to what extent the activity poses a reasonable or unreasonable risk to human dignity. Activities that pose unreasonable risks should be banned, but activities that pose reasonable risks may be permitted with some regulation or guidance. While it is widely held that it is wrong to intentionally, negligently, or recklessly harm people, it is not wrong to create a reasonable risk of harm. Some risks, such as driving an automobile down a sidewalk or firing a gun into a crowd, are so unreasonable that we consider them immoral (and often illegal). Other risks, such as driving an automobile (normally) or operating a gas furnace, are considered reasonable. What's the difference between a reasonable and an unreasonable risk? According to a common view (Feinberg 1973), the difference depends on several factors:[6]

• The probability and magnitude of the harms;
• The probability and magnitude of the benefits;
• The basic rights at stake, such the right to liberty, free speech, free practice of religion, and privacy.

To decide whether a risk is unreasonable, one must weigh these factors. If the probability and magnitude of the potential harm to the potential victim outweighs the benefits to the agent and his or her rights, then the risk is unreasonable. For example, automobiles can cause significant harms to potential victims but they also have important benefits to automobiles users. Moreover, they play a fundamental role in enabling people to realize their rights to education, work, and healthcare. Thus, driving is generally a reasonable risk, but we impose a variety of regulations on this activity in order to reduce the risk. Some types of driving may be judged to pose an unreasonable risk to others and we may decide to forbid them to reduce the risk. For example, drag racing creates tremendous risks, provides questionable benefits, and does not involve fundamental rights. The activity poses an unreasonable risk and we may take actions to prevent it.

INSIGHTS FROM DECISION THEORY

To obtain a better understanding of what makes a risk unreasonable, it will be useful to discuss some basic ideas from decision theory. Decision theory is a mathematical approach to decision making used to represent choices in economics, political science, computer science, psychology, philosophy, and evolutionary biology. Decision theorists distinguish between two types of decisions we make when faced with uncertainty, decisions under risk and decisions under ignorance (Resnik 1987). Decisions under risk are decisions where we can evaluate the expected utility of various outcomes, given our estimates of the probability that particular states of the world will obtain. For example, in deciding whether to take an umbrella to work, my choices would be to take an umbrella or not take an umbrella. The states of the world are it rains or it doesn't rain. My outcomes could be, I don't get wet and have to carry an umbrella (01), I get wet and have to carry an umbrella (02), or I get wet and don't carry and umbrella (O3) and I don't get wet and don't carry an umbrella (04). Suppose that I slightly prefer to not take an umbrella, so I assign this a utility of –2, and that I strongly prefer to not get wet, so I assign this a utility of –10. I prefer going to work without an umbrella somewhat, so I assign this a value of 2. The probability that it will rain is 50 percent. So, I multiply all of my values (magnitudes of benefit or harm) by the probabilities to get expected utilities. I can sum up these expected utilities to assess our decision using the following matrix:

	It rains	It doesn't rain	Sum
I carry an umbrella	$-2 \times .5 = -.1$	$-2 \times .1 = -.5$	-2
I don't carry an umbrella	$-10 \times .5 = -5$	$2 \times .5 = 1$	-6

The most reasonable choice, on this scheme, is the choice that maximizes my expected utilities, which would be the choice to carry an umbrella. I would be taking an unreasonable risk if I did not take the umbrella to work. If the chance of rain were only 10 percent, then the situation would be much different. In this case the sum of expected utilities for carrying an umbrella would be: $-.2 + -1.8 = -2$. But the sum for not carrying an umbrella would be: $-1 + 1.8. = .8$. So, in this case the most reasonable choice would be to not carry an umbrella, and I would be taking a reasonable risk by not taking the umbrella to work.

This approach to decision making assumes that we can assign probabilities to various outcomes based on scientific evidence. Sometimes in formulating public policy we can assign probabilities to various outcomes. For example, in deciding whether to allow a new drug to be sold on the market, the Food and Drug Administration (FDA) obtains scientific evidence to support probable benefits and harms to consumers. In evaluating the environmental impact of a pesticide, the Environmental Protection Agency (EPA) can obtain evidence to calculate potential harms to the environment or people. The disciplines of cost-benefit analysis and risk management evaluate policies based on estimates of the probability that particular outcomes will occur (Shrader-Frechette 1991).

But suppose we lack probability estimates. Suppose that we do not know, for example, the probability of rain. We have no scientific evidence about whether or not it will rain and we cannot even make a reasonable guess about the probability of rain. How should we make a decision in this case? This would be a decision under ignorance. Our new decision matrix would be:

	It rains	It doesn't rain
I carry an umbrella	− 2	−2
I don't carry an umbrella	−10	2

What is the most reasonable thing to do under these conditions? What is a reasonable risk? According to the principle of insufficient reason, one could simply assume that all states of the world are equally probable and just sum up the utilities for each choice to get the option with the highest utility. So, the utility for carry an umbrella would be −4 and the utility for don't carry an umbrella would be −8. So, I shouldn't carry an umbrella. But suppose one of our outcomes is very undesirable. Suppose I am deciding whether to fire a gun and I don't know whether it is loaded and one of the outcomes is that I kill someone. In this case, the most reasonable choice would be to avoid the worst possible outcome. I would be unreasonable to fire the gun. According to this decision rule, known as the maximin rule, we should make the choice that avoids that worst possible outcome or the worst-case scenario. So, we should not fire the gun and we should always take an umbrella (Resnik 1987).

On the other hand, we might also have an outcome that is very desirable and we may not want to miss out on this opportunity. Suppose we are trying to decide whether to test out a new invention, such as an automobile or airplane. The maximin rule would instruct us to not test out the invention, since we might crash. It is a rule for people who are pessimists (they think something bad will happen) and risk-aversive (they want to avoid bad outcomes rather than seek good ones). Someone who is a bit more of an optimist and an opportunist might adopt the minimax regret rule, which holds that we should make the choice whose maximum regret is minimal, where regret is a function of missed opportunities as well as bad outcomes.

What is the best rule to follow for decisions made under conditions of ignorance? A great deal hinges on our attitudes toward risks and opportunities. Are we pessimistic/risk-aversive, optimistic/risk-seeking, or somewhere in between? We must make many important personal decisions, such as where to go to college, whom to marry, or what career path to take, despite knowing very little about what is likely to happen under each decision scenario. We must also make many important public policy decisions, such as whether or not to sign a treaty, fund a new space station, or go to war, in the face of considerable ignorance. We would like to have a great deal of information when we make important decisions, but we often lack enough information to even assign probabilities to various outcomes.

One might argue that we should put off all decisions made under conditions of ignorance until we have enough evidence to calculate probabilities, thereby converting them into decisions under risk. Sometimes this is possible. For example, when buying a new house or car, an individual may be able to wait for sufficient information to make a choice that maximizes expected utility. Likewise, the FDA may be able to wait for enough data before approving a new drug. However, this is often not the case. How should we proceed when we do not have enough time to wait for the evidence we require?

THE PRECAUTIONARY PRINCIPLE

One influential approach to this decision problem is known as the Precautionary Principle (PP). Environmentalists, policy analysts, and politicians have invoked the PP to defend a variety of policies designed to protect public health and the environment, such as restrictions of greenhouse gases or electromagnetic fields (Foster, Vecchia, and Repacholi 2000). There are many different versions of the PP, ranging from extremely risk-aversive to less risk-aversive versions (Cranor 2001). According to one version of PP, we should not allow scientific uncertainty to undermine our ability to respond reasonably to plausible threats to the environment or society. The PP urges us to take reasonable precautions to avoid plausible threats in the face of factual uncertainty (European Commission 2000). It does not instruct us to take all conceivable precautions or to address all worst-case scenarios; precautions should be reasonable and worst-cases scenarios must be plausible (Resnik 2003a).

There is a difference between a *probable* threat, *plausible* threat, and a *possible* threat. A threat is probable if we have sufficient evidence to assign it an objective probability (i.e., a probability based on statistical inferences from observational data or a probability based on logical analysis).[7] For example, based on epidemiological data, we can say that a woman has roughly a 10 percent probability of developing breast cancer by age seventy. Breast cancer is therefore a probable threat. However, we lack sufficient data to determine a woman's chance of being killed by a giant squid. As far as we know, there are no reported deaths of a woman being killed by a giant squid. But we know that giant squids exist, that they can come to the surface of the ocean, and that they are carnivorous. Hence, the threat is possible. But is it plausible? To determine whether the threat is plausible, we must consider whatever data we have that are relevant to the hypothesis and appeal to epistemological criteria, such as simplicity, fruitfulness, precision, and explanatory power, to form a judgment about the plausibility of the hypothesis. The hypothesis is plausible if we decide that these evidentiary considerations justify further attempts to test the hypothesis. Thus, a plausible hypothesis is a live option that deserves further scrutiny and inquiry (Resnik 2003a).

What makes a response reasonable? A reasonable response is a response that a reasonable person would take.[8] There is, of course, no such person. The reasonable person is a legal and philosophical fiction used to draw inferences in law and social policy (*Black's Law Dictionary* 1999). For example, in tort law the standard of care frequently used in a negligence action is the reasonably prudent person standard (Diamond, Levine, and Madden 2000). Many feminists argue that the courts should use the reasonable woman standard in sexual harassment cases (Swisher 1995). The best way to think of the reasonable person is that he or she is a person who makes decisions based on various normative ideals for practical reasoning. Some of the commonly discussed ideals are as follows:

- The reasonable person weighs risk, benefits, opportunities, and costs prudently;
- The reasonable person takes effective means to his or her goals;
- The reasonable person takes actions that are proportional to potential results: he or she does not use a sledgehammer to crack an egg;
- The reasonable person takes a realistic attitude toward his or her situation: he or she does not entertain Pollyannish fantasies or paranoid delusions.
- The reasonable person strives for consistency and economy in his or her conduct;
- The reasonable person obeys the law.

The reasonable person need not follow all of these norms in all decisions. Moreover, there may be norms for reasonableness other than the ones I have mentioned here. These norms should therefore be treated as useful desiderata, but not as necessary and sufficient conditions for defining reasonableness.

An example will help to illustrate how one might apply the PP. Consider the precautions Jane Doe might take related to driving to meet her appointment for a job interview in a town thirty miles away. Consider the hypothesis that Jane will have a flat tire. Having a flat tire is always a plausible threat when driving, even though we cannot assign an objective probability to the hypothesis that Jane will have a flat. Flat tires result from so many different factors, such as ordinary wear and tear, product defects, and road hazards, that it is impossible to predict whether any particular car will have a flat tire on any given day. Nevertheless, Jane has some basis for believing that a flat could happen to her and for preparing for this scenario. Consider the hypothesis that a meteor will hit Jane's car. This threat, though possible, is not even plausible. There is just not enough evidence to support this hypothesis. So Jane should not waste her time preparing for this threat. What would be a reasonable response to the threat of a flat tire? The PP would advise Jane to take a spare tire while driving as a reasonable precaution against a flat tire. This response is prudent, proportional, realistic, effective, and economical. Other responses would be unreasonable. For example, driving without a spare would be unreasonable because it would be not be prudent, realistic, effective, or economical. Staying at home and not driving would also be unreasonable because it would not be prudent, realistic, economical, or proportional.

A principle like the PP could have wide applications in science and technology policy, since we must often make decisions about the development of science and technology despite considerable uncertainty. People have appealed to the PP in debates about global warming, pollution, genetically-modified crops, and public health. I shall refer to this principle again in this book, since I think it also may lend some insight into the DNA patenting debate.

Turning from this abstract discussion to practical questions about DNA patenting, we can think of threats to human dignity (as well as other threats posed by DNA patents) as potential harms that we must evaluate under conditions of uncertainty. To determine whether DNA patents pose a reasonable or unreasonable risk (or justified or unjustifiable threat), we should approach this question with the best evidence we have at our disposal as well as insights from decision theory. Later on in this chapter I will evaluate the threat to human dignity posed by DNA patents, but first I will consider the prior question of whether DNA patents violate human dignity, since if these patents violate human dignity, they would be unethical in any case.

DO DNA PATENTS VIOLATE HUMAN DIGNITY?

Based on the discussion earlier, we can say that DNA patents would violate human dignity if they treated people as complete commodities. But do DNA patents treat people as complete commodities? To answer this question, let us consider the example of slavery, which is a practice that treats people as complete commodities. The slave owner has absolute dominion and control over the slave just as an automobile owner has complete dominion and control over his car (*Black's Law Dictionary* 1999). He has the right to sell the slave, to compel the slave to work, to breed the slave, to rent the slave, and so on. Slavery is dehumanizing because it treats people as chattel; it is the paradigmatic case of a violation of human dignity. Slavery has existed in some form or another for thousands of years of human history and it is still practiced in some parts of the world today. The United States fought a civil war over the issue of slavery and still struggles with the legacy of slavery. Obviously, slavery is a moral abomination that should be avoided. But would DNA patenting be anything like slavery?

There are at least three arguments why DNA patenting is different from slavery. First, as we saw in chapter 3, patent rights are not as extensive as full ownership rights. Patenting only involves the right to exclude others from making, using, or selling an invention; it does not imply positive rights to make, use, or commercialize the invention because exercising these rights may be illegal. DNA patents do not give the patent holder the right to use, make, or commercialize human beings because all of these activities are currently illegal. A DNA patent holder would not hold any positive rights over a human being. At most, he would have the right to exclude others from making, using, or commercializing the human being, but no one would have these rights in any case.

Second, as argued in chapter 5, DNA patents apply to artificial body parts, not to natural ones. If we hold that DNA patents only give the patent holder control over isolated and purified DNA, and that none of the DNA in a human being is in an isolated and purified form, then the patent holder has no rights at all over the DNA contained in a living, human being. DNA in human beings exists in its natural state, but, as argued in chapter 5, DNA patents should only apply to artificial DNA.

Third, DNA patents apply only to parts of human beings, not to whole human beings. The human body has many different parts, ranging from organs, tissues and cells, to proteins, lipids, hormones, and minerals. As we saw in chapter 2, DNA is a body part. It is a very important body part, but a body part nonetheless. A patent on DNA is not a patent on a whole human being but only a patent on a human body part (Resnik 1997).

These three arguments demonstrate that DNA patenting is not equivalent to the complete commodification of whole human beings found in slavery. Only a patent on a whole human being would imply complete commodification of a whole human being. A patent on a whole human being would violate human dignity and would be immoral.[9] DNA patents apply market rhetoric to human body parts, not to whole human beings. Moreover, the body parts are arguably invented body parts. Thus, patenting DNA is more like patenting an artificial heart valve or hip than it is like patenting a human being. If patents on artificial heart valves do not treat human beings as complete commodities, then patents on DNA also do not treat people as complete commodities (Resnik 1997).

On the other hand, one might challenge these arguments. Let's consider some possible objections.

One might agree that patent rights are not as extensive as full ownership rights yet still claim that this assertion misses the main point, because patents do involve some type of commodification, which is the main issue (Hanson 1997, 1999). A person who rents a house has less dominion and control over that house than a person who owns the house, but we would still say that renting is a type of commodification (or commercialization). Likewise, a person who patents a whole human being has less dominion and control over that human being than a person who owns a human being, yet we would still say that patenting does treat the whole human being as a commodity. Although I think we have good reasons to believe that patenting does not constitute complete commodification per se, I will grant this point. Even if patenting is not the same thing as slavery, patenting a human being would be a morally disturbing type of commodification and would violate the dignity of that human being.

One might admit that, technically speaking, DNA patents apply only to artificial body parts yet also argue that the relationship between artificial and natural DNA is so close, that patents on artificial DNA constitute commodification of natural DNA. As we saw in chapter 2, artificial DNA is derived from natural DNA via chemical and biological processes. Very often, the homology between natural and artificial DNA is as high as 95 percent, meaning that 95 percent of the sequences found in the artificial sequence are also found in the natural one. Thus, the relationship between natural DNA and artificial DNA is like the relationship between a newspaper article and a photocopy of the article. The article as it exists, in the newspaper, is not in an isolated or purified form. The copied article is an isolated version of the article. One might argue that some actions one performs on the copied article also imply conduct toward the original. For example, reading the copy provides the same information as reading the original. If you have read the copy, then you have, for all practical purposes, read the original. One might also argue that quoting from the copy is quoting from the original. Some of the same points apply to DNA. Making a protein from an artificial sequence is like making a protein from its corresponding natural sequence. Taking a splice from an artificial sequence is like taking a splice from its corresponding natural sequence.

But the analogies only go so far. A person who owns a photocopy of an article does not own the article. A person who burns a copy of an article has not burned the article. Likewise, a person who has property rights over an artificial DNA sequence does not, therefore, have property rights over its corresponding natural sequence. I apologize for belaboring this point, but I will state it again: a person who patents an isolated and purified DNA sequence does not obtain (or should not obtain) any property rights over its naturally occurring counterpart. Ergo, commodification of artificial DNA does not imply commodification of natural DNA.

The objector has one more argument up his sleeve. The objector could argue that the actions performed on artificial DNA may symbolize actions performed on natural DNA. For example, consider burning an effigy of a person. Since the effigy is a symbol of the person, burning the effigy makes a symbolic statement about the person. The act of burning the effigy defiles or denigrates the person. Likewise, one might argue, commodifying artificial DNA makes a symbolic statement about natural DNA (e.g., that it is a commodity and has only commercial value). In the same way, the act of commercializing a symbol often makes important statements about the value of the thing that is symbolized.

On the other hand, from the mere fact that we commercialize a symbol, we cannot infer that we therefore believe the thing that is symbolized has only a market value. For example, we sell crosses and various religious symbols, yet most people would not say that this means that we believe that God has only a market value. We also commercialize American flags, portraits of George Washington, and photos of Niagara Falls. It is quite reasonable to hold that while the symbols have a market value, the things symbolized still have a nonmarket value. There is, of course, some danger that the act of commercializing a symbol will encourage us to view the thing symbolized as having only a market value. Although selling the American flag is not the same thing as selling America, it could encourage us to view America as something that is for sale. This is an interesting argument that I will return to in this book. I will note, however, that since it focuses on the harmful consequences of treating a symbol as commodity, it does not constitute a deontological argument against such a practice. Thus, this argument could provide more data for some overall utilitarian analysis of DNA patents in that it would have us consider the harmful consequences of patenting something that symbolizes natural DNA.

PARTS VERSUS WHOLES

In objecting to the previously mentioned third argument, one might admit that there is a distinction between parts and wholes but argue that sometimes commodification of a part of a thing implies commodification of the whole thing. To make this sort of counterargument, one must be careful not to run afoul of various part/whole fallacies and confusions (Gorovitz et al. 1979). Inferences from properties of the parts of a thing to the properties of a whole thing are deductively invalid (Copi 1986). From the fact that a part of a car (e.g., a tire) is made out of rubber, one cannot infer the whole car is made out of rubber. From the fact that a part of a field is barren, one cannot infer that the whole field is barren. Thus, from the fact that a part of the human body has been commodified, one cannot infer that the whole human body has been commodified (Resnik 1997; Resnik 2001d).

To get around this problem, one might claim that part/whole inferences, while deductively invalid and generally unwarranted, can be justified when there are appropriate connections between the parts and the whole. For example, if a player cheats on one move in a game of chess, then they have cheated on the whole game. If a radio station airs two minutes of commercials every hour, then the whole broadcast is commercial radio. If the foundation of a large building is destroyed, the entire building will be destroyed. If one part of a building has ugly graffiti on it, the entire building may be ugly. If an otherwise virtuous person commits an egregious crime, we may regard the whole person as vicious. In thinking about these connections between parts and wholes, we can classify them as follows:

Conceptual Connections: Sometimes properties of a part are conceptually connected to the whole. For example, a game of chess is defined by a series of moves and an illegal move therefore affects the legality of the whole game.

Causal Connections: Some parts and wholes are causally connected. For example, a foundation is causally connected to the rest of the building structure.

Axiological (or value) connections: Sometimes moral or aesthetic value judgments we make about parts affect our judgments relating to the whole. For example, our judgment of the aesthetic quality of a part of a building may affect our judgment of the quality of the whole building. Our judgment of concerning one particular law in society may affect our moral assessment of the whole society. For instance, one might argue that a society that permits slavery is unjust even if has many other morally redeeming features.

Our question then becomes the following: how might the commodification of a part of the body be connected to the whole body? In terms of the three types of connections mentioned above, the last one seems to be the best way of approaching these questions, since we are concerned about how our value judgments relating to the parts affect our value judgments concerning the whole. We can restate the question as follows: if we regard a part of the body as having only a market value, does this imply that we also view the whole body itself as having only a market value? This is the question we tried to answer earlier while probing the depths of Kant's philosophy.

Perhaps our answer to this question depends on what type of body part we have in mind, since, as we have seen, there any many types of body parts. For example, the body contains basic elements, such as carbon, sulfur, nitrogen, hydrogen, oxygen, and calcium. The body also contains a variety of molecules ranging in size from water and glycogen to hemoglobin and DNA. It would be absurd to suggest that buying and selling sulfur or water amounts to commodification of the human body (i.e., treating a person as property), even though sulfur and water are body parts. Likewise, one might argue that buying and selling proteins, insulin, hair, sperm, blood products, teeth, and even cell lines does not imply commodification of the whole human body (Resnik 2001d). But what about selling vital organs, such as the brain, the heart or the lungs? Are there some body parts that should not be commodified at all because treating those parts as commodities is the same thing as treating the whole body as a commodity? Where can we draw the line when it comes to commodifying the body? We saw how Kant struggled with this question; we are still struggling with it today.

I will not explore the issue of commodifying human organs in this book (see Andrews and Nelkin 2001 and Gill and Sade 2002 for further discussion), but I will observe that DNA is not the same as a human organ. DNA is a complex macromolecule; it is not a highly developed structure like a heart or kidney. The body contains billions of copies of DNA but only one liver or heart. DNA can regenerate itself; most organs cannot regenerate. DNA is not an essential body part, since a person can lose some of their DNA and not die. Many organs, on the other hand, are essential. Viewed in this light, commodification of DNA should be no different from commodification of proteins or other large macromolecules in the human body. If treating proteins as commodities does not imply that the body is a complete commodity, then treating DNA as a commodity should not imply that the body is a complete commodity. If protein patents do not violate human dignity, then DNA patents should also not violate human dignity.

DNA AND GENETIC IDENTITY

But one might reply that DNA is no ordinary macromolecule because it contains the genetic information required to make a human being. With the exception of identical twins, everyone has a unique genome. Although DNA is not an essential body part, it is a special body part closely connected to the person (Brody 1999; Hoedemaekers and Dekkers 2002). Due to its special status as the blueprint for a human being and its ability to serve as a genetic fingerprint, treating DNA as a complete commodity is like treating a person as a complete commodity: if a person's DNA is for sale, then that person is for sale. There is an important axiological connection between DNA and the whole person. Thus, patenting a person's DNA violates the dignity of that person.

To evaluate this argument, we need to say a bit more about the connection between genetic identity and personal identity, since this argument asserts that a person's DNA has some axiological connection to the person. There are two basic approaches to the problem of personal identity, the physiological approach and the psychological approach. According to the physiological approach, a person is identical to his or her body. The person exists from the time he has a distinct living body until his body dies. Although there are some disputes about when this historical sequence begins (e.g., at conception), the emergence of the primitive streak, viability, birth, or at some later time, proponents of the physiological approach all hold that personal identity begins and ends with the existence of the person's distinct, living body. From an objective point of view, this approach makes a great deal of sense in that we identify the people we know through references to their bodies. To distinguish between two different people, we use physical traits, such as eye color, hair color, height, weight, voice, and fingerprints.

There are some difficulties with this approach, however. First, where in the body is the person located? Is there some part of the body that is more essential to the person than some other part? For example, if someone survived a terrible car accident and had burns over 80 percent of their body with partial paralysis and complete blindness, we would probably say that they looked like a different person. Indeed, their body would be very different. But they would still be the same person even if their body had changed. To make sense of this intuition, one must say that the person cannot be equated with his hair, skin, eyes, legs, or other body part that would be destroyed in such an accident.

Second, many would argue that it is possible for a body to remain the same and for the person to change or die. Suppose that someone has an automobile accident and becomes brain dead; they have no brain activity. The body is still alive—machines are supporting its functions—but is the person gone? Under the widely used whole brain definition of death, death occurs when the whole brain has lost permanently its various functions (President's Commission 1981). If death can be legally declared as this point, then legally speaking (and I would also say morally speaking), the person no longer exists, since taking vital organs from a brain dead patient does not constitute murder. To donate a vital organ, a person must already be dead (*In re T.A.C.P.* 1992).

To deal with these and other problems with personal identity, many writers have defended the psychological approach, which holds that the person can be identified with their memories, character traits, interests and likes, fears and wants, cognitive skills, emotional disposition, attitudes and feelings, personality, and other aspects of their mind. This approach explains how someone could be the same person before and after a terrible automobile wreck: their body has changed but their mind remains the same. It also explains the moral basis for using the whole brain definition of death: when a person's mind no longer functions properly, the person is dead.

On the other hand, the psychological approach has its own difficulties. First, the approach would seem to imply that someone who radically alters their beliefs, attitudes, or personality is no longer the same person: a drug-abusing criminal who reforms and becomes a born-again Christian might not be the same person, under this approach. Many people would want to say that he is the same person even though his personality has changed. Second, this approach would also imply that someone who is not brain dead but still loses a great deal of mental functioning is, in a sense, dead. Many people would want to say that a person with higher brain death as found in a persistent vegetative state or end-stage Alzheimer's dementia has not yet died, even though most of that person's mind is gone (Bernat, Culver, and Gert 1982). Third, the psychological approach would seem to imply some type of metaphysical dualism: if the person cannot be equated with some particular body part, such as the brain, then where is the person? Could the person move from one body to another body? How could a person be in the body if one cannot locate the person in the body?

In response to these, as well as other concerns, some writers have argued that personal identity cannot be equated with physical traits of the body or mental traits of the mind, but consists in the unique combination of physiological and psychological traits (see Hamlyn 1984 and Parens 1996 for further discussion). A person is neither a mind nor a body but a unique combination the person's whole mind and whole body. This view implies that a person cannot be equated with any particular body part: amputating a leg is not equivalent to killing the person. A person who has burns over 80 percent of their body is still the same person. On the other hand, since this position rejects metaphysical dualism, it accepts the idea that there is some important connection between the person and his or her body. For example, destroying a person's body, or taking an action that would lead to the destruction of the body, would destroy the person. Thus, this view would object to removing vital organs, such as the heart or brain, from the person's living body, because such removal would lead to the destruction of the person. The view would also object to the commodification or removal of body parts that bear an important connection to the person. But what body parts have such as connection? Does the human genome have an important connection to the person?

There has been a great deal of confusion about relationship between the person and his or her DNA in recent years, as some people have claimed that the genome is the human soul, while others have wondered whether a cloned individual would even have a soul (because they would lack a unique genome). These confusions reflect widespread infatuation with the idea that personal identity is genetic identity. We should reject this naïve genetic reductionism. Personal identity is not the same as genetic identity (Nelkin and Lindee 1995). Although the genome plays a very important role in the production of physiological and behavioral phenotypes, it does not causally determine phenotypes. Identical twins have virtually the same genes but have different personalities and cognitive and emotive traits. Identical twins may even have physiological differences due to environmental influences on growth, accidents, illnesses, and so on. There is no scientific basis for the popular myth that a clone of a person would be the same person. A clone of Michael Jordan would not be Michael Jordan and might not even like basketball.

One might admit that human identity should not be equated with genetic identity yet still maintain that the genome is a major component of human identity and bears an important connection to the person (Brody 1999). Although a person cannot be equated with a particular genome, the genome can be causally and axiologically connected to the person. The genome has a causal connection in two respects. First, the existence of a person's genome is a necessary condition for the existence of that person, if my genome had never been formed, then I would not exist. Second, the genome plays a significant causal role in the production of the phenotypes that we use to identify the person, such as their height, weight, eye color, personality, and so on.

The genome also has an axiological connection to the person because modern societies, such as the United States, place a great deal of moral and cultural significance on human genetics. People seek genetic explanations of human behavior and social phenomena and genetic interpretations of responsibility and guilt. Since the gene has become an important cultural icon, what we say about a person's genome has important implications for our moral assessment of that the person (Nelkin and Lindee 1995). Whether we like it our not, many people blame criminal behavior on bad genes and associate success with good genes.

We can now develop a moral argument against patenting the genome based on relationship between the genome and the person. Because the human genome has strong causal and axiological connections to the human person, commodifying the human genome violates human dignity (Brody 1999). Since patenting DNA commodifies the human genome, it violates human dignity. Thus, patenting DNA is immoral.

I think this is a strong deontological argument for placing moral restrictions on the commodification of the human genome, since commodifying something that bears an intimate connection to a person's identity is almost like commodifying the person. A person who held a patent on a whole genome would hold a patent on material that could be used to generate a human embryo, which could then develop into a human being. (One could create an embryo with a patented genome by using nuclear transfer procedures similar to those used to create Dolly, the world's first cloned sheep.) Thus, commodification of the DNA that composed that human being would imply commodification of that human being. Therefore, patenting a whole human genome would violate human dignity. The genome has special status as a body part, and should be treated accordingly. Although commodification of a whole human genome would fall short of slavery, it would be morally disturbing and would warrant some form of prohibition. For these reasons, I believe that it would be wrong as a matter of principal to patent a whole human genome. Patenting a whole human genome would be immoral, regardless of the consequences (Resnik 2001b).[10]

On the other hand, the intimate connection between the person and their DNA only holds between the whole genome and the person; it does not hold between parts of the genome and the person. Patenting DNA sequences, gene fragments or even a set of genes would not violate human dignity because these smaller parts of the genome do not bear a special causal or axiological relationship to the person. One cannot use nuclear transfer techniques to create an embryo from a gene. Also, we do not attach the same moral and cultural significance to individual genes or gene fragments that we do to whole genomes. Therefore, the patenting of DNA sequences, gene fragments, and the like, is not inherently wrong; the wrongfulness of these activities depends on their consequences for science, medicine, agriculture, and society. Commodification of parts of a human genome would constitute only incomplete commodification of a human being. As such, patents on parts of the genome may pose a threat to human dignity, but they do not violate human dignity. We have now answered Kantian concerns about patenting the human genome: we can draw the line at the whole genome. Patenting a whole human genome would treat a person as having only extrinsic value.

The European Commission (1998) has declared that patents on procedures used to modify the human germ-line would violate human dignity. I believe that such procedures may threaten, but do not, violate human dignity. In declaring such patents immoral, the European Commission has expressed moral objections to the genetic engineering of human beings for eugenic or other nontherapeutic reasons. Germ-line modifications raise a variety of moral issues, including risks to progeny and future generations, risks to the gene pool, justice, discrimination, effects on people with disabilities, and the question of enhancement (Resnik, Steinkraus, and Langer 1999). Most of the moral questions relating to germ-line modification have to do with the application of technologies, not to the patenting of technologies per se. Techniques or procedures that could be used to engineer people for eugenic purposes could also be used to prevent or treat genetic diseases. Rather than explore all of these issues here, I will suggest that most of these concerns identify threats to human dignity but not violations of human dignity. Moreover, violations of human dignity relating to patents on germ-line modification technologies would arise only if any of these patents treat whole human beings (or whole human genomes) as mere commodities. Since most of the patents related to germ-line modifications would involve technologies designed to modify only parts of the genome, they would not violate human dignity.

Even if we do not allow patents on a whole human genome, wouldn't patents on all or most of the bits and pieces of the genome constitute complete commodification of the genome and therefore violate human dignity? One might argue that since a whole is the sum of its parts, treating all the parts of a human genome as commodities is equivalent to treating the whole human genome as a commodity (Nelkin and Andrews 1998). Although this point raises some interesting concerns, its conclusion does not follow from its premises. While the genome is composed of its parts, it is not merely a sum of its parts. Patenting parts of the genome is not equivalent (legally or morally) to patenting a whole genome. For comparison, the different parts of an automobile may be patented, but this does not imply that the whole automobile itself is patented. A human body may contain many patented parts, ranging from proteins and DNA, to artificial hips and ears. But the patenting of these parts, individually, would still not imply a patent on the whole.

In any event, it is likely that a high percentage of the human genome will be in the public domain because publicly and privately funded research programs are publishing the human genome in public electronic databases, and the PTO has raised the bar on the utility requirements for DNA patents. Inventors will still be able to obtain patents pertaining to specific uses of DNA, such as patents on the use of DNA in diagnosis or drug development, but researchers will not be able to patent thousands of genes or gene fragments in an attempt to stake out a claim on the genome. Inventors will be able to patent only those parts of the genome that have definite uses.

To summarize this section briefly: I have argued that the vast majority of DNA patents do not violate human dignity, although they may threaten it. To violate human dignity, a patent must treat a person as a complete commodity. Most DNA patents treat human beings as incomplete commodities. Only a patent on a whole human genome would treat a whole human being as a complete commodity. So how should we think about the potential threats to human dignity posed by DNA patents that do not assert claims on a whole human genome? I will now tackle that question.

Do DNA Patents Threaten Human Dignity?

Most of the human dignity concerns discussed in this chapter have exposed potential threats to human dignity posed by DNA patenting. As such, these concerns constitute a type of slippery slope argument against DNA patents. The argument proposes that treating human beings as incomplete commodities will lead to further commodification of human beings, as well as other undesirable outcomes associated with commodification, such as exploitation. According to this argument, market values will corrupt or drive out all other values that we might place on human beings. At the end of this precipitous slide, we will no longer respect or value persons, and we will regard human beings as commodities or chattel (Kimbrell 1997; President's Council 2002). It is worth noting that slippery slope arguments have been proposed in many different areas of debate in bioethics. For example:

- Some people have argued against voluntary physician-assisted suicide on the grounds that it will lead us down a slippery slope toward euthanasia, involuntary assistance in death, unjust discrimination against the disabled, genocide, and others horrors (Smith 2001).
- Some people have argued against allowing patients or families to sell blood, tissues, and organs because commercialization of the body will create a slippery slope that will crowd out and destroy voluntary donation of blood, tissues, and organs (Murray 1986).
- Some people have argued against the genetic modification of the human genome to prevent diseases on the grounds that this practice will lead us toward genetic enhancement (Kass 1985; Rifkin 1983).

- Some people have argued against surrogate pregnancy contracts on the grounds that they will lead us down a slippery slope toward increasing commodification of women and children and the exploitation of women (Anderson 1990).
- Some people have argued against prenatal genetic testing on the grounds that it will lead us toward discrimination and eugenics (Rifkin 1983, 1985; Kass 1985).
- Some have argued against in vitro fertilization on the grounds that it will send us down a slippery slope toward the destruction of traditional families and marriages (Kass 1985).
- The President's Council on Bioethics (2002) developed a variety of slippery slope arguments against reproductive and research cloning, including the arguments that research cloning will lead to reproductive cloning and that reproductive cloning will lead to eugenics, the commercialization of procreation, troubled family relations, and loss of respect for human life.

Slippery slope arguments have a great deal of rhetorical and emotional force, which is one reason why people use them so often in public debates. Some slippery slope arguments are good arguments, others are poor arguments. To distinguish between good and bad arguments, we need to first remind ourselves that all slippery slope arguments are deductively invalid and that at best they can offer inductive proof for their conclusions (Kahane 1990). Thus, anyone who relies on a slippery slope argument to support a policy conclusion must be careful to provide empirical evidence for that conclusion and to assess the strength of that evidence. Inductive arguments, though deductively invalid, may be regarded as stronger or weaker, depending on the quality and quantity of the evidence proffered (Copi 1986). For example, consider the following slippery slope argument we might make to a high school senior. "You are doing very well in life at the moment, so stay away from marijuana. If you try marijuana, you will also try other drugs. If you try other drugs, you will become an addict. If you become an addict, you will drop out of school and wind up in jail or dead." This argument presents the person with a long chain of if-then clauses that lead to some undesirable results. To evaluate this argument, we must assess all of the links in the chain to see if they are likely to lead to the results. To determine the inductive strength of a slippery slope argument, we must examine each link in the chain and ask ourselves whether there are some steps we can take to stop the slide. For example, maybe our student can stop the slide from marijuana to other drugs. Or maybe he can stop the slide from drug usage to drug addiction. Or maybe he can stay in school.

It is notoriously easy to tell some kind of a story that leads from an event to some dreadful outcome such as the end of the world. Like Chicken Little, all one needs is a bit of paranoia and an active imagination: "If you chew that sour gum, you will spit it out immediately. If you spit it out immediately, Dr. Raffiki will see you. If Dr. Raffiki sees you, he will go insane. If he goes insane, he will help terrorists obtain nuclear materials. If terrorists obtain nuclear materials, then they will build nuclear weapons. If they build nuclear weapons, they will use them on both the United States and Russia to get both sides to believe that the other is at fault. If both the United States and Russia believe that they have been attacked, they will retaliate. If they retaliate, they will detonate enough nuclear weapons to destroy the whole world. So don't chew that sour gum!" The problem with this fanciful example is that the story cannot even be assessed as an inductive argument because we have no basis for evaluating its premises. It is simply an argument that postulates a chain of events that could lead to a bad result. I believe that many slippery slope arguments tossed around casually in public debates are no better than this story about chewing sour gum, because they offer no credible evidence connecting the first part of the chain to the disastrous result.

If we take a close look at our lives, we will see that we are surrounded by many potential slippery slopes. There is virtually no way to steer clear of all potential slippery slopes; we must take some risks in order to live and accomplish our goals. The moral life does not consist in avoiding all potential harms, but rather it consists in making wise choices about benefits, harms, risks, and opportunities. Since slippery slopes involve the possibility of some bad outcome, a reasonable person will assess each slippery slope according to the evidence and take reasonable risks.

The first step in assessing a slippery slope argument is to determine whether it can be evaluated on inductive grounds: What is the probability of the conclusion, given the argument's premises? Are the premises proven? Are they speculative? Are the premises true, improbable, or highly probable? If we can assign the conclusion some degree of probability, then we can treat decisions relating to the slippery slope argument as decisions under risk, and we can take actions designed to maximize expected utilities. For example, we know that a high percentage of people in society will become alcoholics if they are allowed to drink alcohol. For some people, there is a slippery slope from alcohol use to alcoholism. In response to this problem, we could choose among several options for the commercialization of alcohol.

Option 1: Sell alcohol to anyone. Do not regulate it.

Option 2: Forbid the sale of alcohol. Punish people who sell it.

Option 3: Regulate the sale of alcohol.

In order to decide which is the best option, we can follow the method outlined earlier in this chapter. We can examine the benefits and risks of each option as well as any relevant rights that may be at stake, such as the right to autonomy. Although alcohol poses a threat to human health, we might decide against paternalistic bans on alcohol in order to allow some people to enjoy the benefits of drinking and to exercise the autonomy to choose whether or not to drink. The risks of alcohol and its threats to human health can be justified based on its benefits to people and the rights at stake.

We could apply the same method to DNA patents as well and treat decisions about DNA patents as decisions under risk. In deciding whether to allow DNA patents we would therefore need to understand:

- The benefits of DNA patents.
- The potential harms of DNA patents.
- The relevant rights at stake in DNA patenting.

The benefits of DNA patents, which we have discussed earlier, include encouraging investment in biotechnology; promoting the advancement of science, medicine, and agriculture; and job creation, economic growth, and so forth. Some of the harms, which we will discuss in more depth later in this book, include harmful effects on the progress of science, medicine, and agriculture. This chapter has addressed potential harms to culture and society stemming from the commodification of DNA. For the most part, these worries focus on the threat of further commodification of human beings posed by DNA patents. If we treat DNA patenting policies as decisions under risk, then we can make a choice that maximizes our expected utilities without violating fundamental rights. Our basic choices would be:

- Ban DNA patents; do not allow them.
- Do nothing; allow DNA patents with no restrictions.
- Allow DNA patents under some form of regulation designed to minimize risks, maximize benefits, and protect human rights.

In this example, it is important to realize that the outcomes depend on the decisions we make; each of these different policies will lead us down different decision pathways. The decision to ban all DNA patents would lead us down a different decision path from the decision to allow DNA patents with no restrictions.

This maximizing approach faces a significant problem: we cannot calculate probabilities relating to the benefits and harms based on different decision pathways. We lack sufficient evidence to make accurate predictions about the consequences of our decisions. Our decisions are therefore decisions under ignorance, not decisions under risk, and we cannot use decision rules that are designed to maximize expected utilities. As we saw earlier, there are several strategies for making decisions under ignorance. If we follow the maximin principle, then we might find that DNA patents raise the possibility of some horrible worst-case scenarios, such as the commercialization of human beings, the privatization of science, and injustices in health care and agriculture (Kimbrell 1997; Rifkin 1998; Shiva 1996). The maximin principle would instruct us to do everything we can to avoid these bad outcomes, including banning DNA patents.

On the other hand, as we saw earlier, one drawback of this principle is that it denies the decision maker (in this case, society) important opportunities. If we ban DNA patents we will miss out on the potential benefits of patenting DNA. Indeed, since new technologies always present potential benefits and harms, the maximin principle is not a very prudent principle for technology assessment. We would still be in the Stone Age if we vowed to take all steps necessary to avoid worst case scenarios associated with new technologies.

I believe our best policy with regard to the threats posed by DNA patents would be to use some form of the Precautionary Principle. If we invoke the PP in this case, then we should take reasonable precautions in response to plausible threat posed by DNA patents. To apply this principle, we need to ask ourselves What are the threats posed by DNA patents? How plausible are these threats? and What are reasonable precautions to take against these threats?

Most of the nightmare scenarios invoked by opponents of DNA patents are not plausible threats. Consider the commonly made argument that commodifying any body part will take us down a slippery slope toward viewing the entire body as a commodity. If we allow DNA to be patented, so the argument goes, we will eventually regard people as chattel or slaves (Kimbrell 1997). Critics of other types of commodification of the body, ranging from contract pregnancy to the selling of gametes and organs, have made exactly this type of slippery slope argument against these other practices (Anderson 1990; Kass 1985).

This threat, that we would regard people as mere property, is indeed a horrible, possible outcome. But is it plausible? A closer examination of the evidence leads us to the conclusion that it is not. First, we have no evidence that the society that allows the greatest amount of commodification of the body, the United States, is edging any closer to laws or policies that would treat people as property. Indeed, the last 150 years of legal history reveals that respect for fundamental rights of all races and both sexes has increased. The United States has eliminated slavery, stopped the subjugation of women, and extended full constitutional rights to all races, ethnicities, and sexes. Although children were once regarded as property, adoption laws prohibit the selling of children. As the notion of treating human beings as persons with fundamental rights has become more prevalent, the view that people are property has receded.

Second, since we currently do not regard human beings as mere commodities despite 300 years of capitalism and incomplete commodification of people, there must be very strong social resistance to complete commodification of people. This resistance enables society to stop the slide toward complete commodification of people. It is not realistic to think that patents on DNA somehow will tip the scale and start us down a slide toward treating people as mere property. Our moral, religious, cultural, legal, and philosophical traditions exert considerable weight on the other side of the scale. Thus, the idea that DNA patenting will lead us down a slippery slope toward treating people as property is pure hyperbole.

While we are assessing slippery slopes related to our treatment of the human body, it is also worth noting that the Catholic Church opposed human dissections for many years. Church doctrine held that the body was the temple of the soul and that dissection would preclude resurrection of the body. Despite the church's opposition to dissections, many physicians, surgeons and anatomists performed dissections of the human body in order to learn more about human anatomy and physiology and to teach their students (Porter 1997). Those who performed dissections frequently obtained human bodies from executed prisoners and some robbed graves. Some scholars have speculated that during the Spanish Inquisition, the great anatomist Andreas Vesalius (1514–1564) was accused of heresy and of dissecting the body of a woman who was still alive, and that he was able to escape with his life only under the protection of Phillip II (Saunders and O'Malley 1982). The taboo against human dissection continued into the eighteenth century and forced many surgeons to rely on grave robbers for cadavers for dissection. Today, society accepts human dissection as a legitimate practice designed to advance human knowledge and medical training. I would venture to say that our acceptance of human dissection has not had a negative effect on our respect for the human body. Accepting this practice has not led us down a precipitous slope toward ghoulishness or cannibalism. If human dissection does not threaten the value we place on human life then surely DNA patenting does not.

Opponents argue that DNA patents, by themselves, do not create a plausible threat to human dignity, but that they do create a threat to human dignity when one considers them in light of all the other practices that commodify the body, such patents on proteins and cell lines, contract pregnancy, and the selling of gametes. All of the various practices that partially commodify the body constitute a commercial juggernaut that will decimate the value we place on human life and will lead us to treat people as mere commodities (Kimbrell 1997; Andrews and Nelkin 2001). My response to this argument is not too different from my earlier responses. I continue to find it to be a significant social fact that we haven't slid down this slope and that we still place great value on the intrinsic worth of human beings despite over 300 years of capitalism and over 200 years of commercial medicine. There are strong social and legal institutions in society that will continue to hold the line against this potential slippery slope. Contract pregnancy and the selling of gametes have not lead us to view children as property, and these practices will probably never have this effect. Some slopes are just not very slippery. These slippery slopes related to DNA patenting are not a plausible threat.

On the other hand, there are some threats to human dignity posed by DNA patents that have a firmer basis in reality. It is possible that DNA patents could encourage the exploitation of human beings for the commercial value of their genes. Exploitation violates human dignity by treating the exploitee as a mere instrument to the ends of the exploiter. The exploiter takes unfair advantage of the exploitee in order to promote his own goals or ends (Wertheimer 1996). For example, a famous legal case involving commodification in cell lines, *Moore v Regents of the University of California* (1990), which we will explore in more depth in chapter 8, illustrates the real risks of exploitation. In this case, Moore argued that he had been commercially exploited for the value of his cell line and that his dignity and autonomy had been violated. In chapter 8 we will consider a couple of other cases involving commercial interests in human DNA that also raise questions of exploitation and injustice. For now, I will observe that the harms addressed in *Moore* represent plausible threats to human dignity: if *Moore* can happen once, then it can happen again, unless we take steps to prevent it.

What would be a reasonable response to the threat of exploitation posed by DNA patents? One response would be to stop the commercialization of human biological materials, but this response would lack prudence, proportionality, and economy. On the other hand, doing nothing to prevent exploitation would also not be reasonable. A reasonable precaution would be to regulate DNA patenting to insure that proprietary control of DNA does not lead to exploitation. For instance, regulatory agencies, such as institutional review boards, the FDA and NIH, and professional societies could require physicians and researchers to inform patients about the potential economic value of tissues, DNA samples, and other biological materials collected during informed consent to research and therapy. These agencies could also require physicians and researchers to offer fair and reasonable financial remuneration to subjects and populations, where appropriate.

Some other potential DNA patents would appear to pose a plausible threat to human dignity even if they do not violate human dignity. Some of these patents include:

- patents on genes associated with personality, intelligence, height, or other desired traits;
- patents on human-animal hybrids (e.g., a "humanzee");
- patents on methods for enhancing the human genome or for eliminating "undesirable" genes, such as genes for homosexuality or alcoholism.

These patents pose plausible threats because they could lead to exploitation, discrimination, or other practices that violate human dignity. As a reasonable precaution, societies may also decide to regulate and perhaps ban some of these patents.

CONCLUSION

This chapter has considered the question of whether human DNA patents violate or threaten human dignity. It has argued that patents on whole human genomes would violate human dignity and should therefore be banned. The vast majority of DNA patents, however, would pose only a threat to human dignity. Some of these threats (i.e., the threat that DNA patents would lead to slavery or other forms of complete commodification of the body), should not be take seriously because they are not at all plausible. Other threats should be taken seriously because they are based on some evidence. The reasonable response to such threats is not to avoid those threats at all costs, but to take precautionary measures to minimize risks posed by those threats. Society can and should regulate DNA patenting practices in order to protect human beings from exploitation, discrimination, and other specific threats to human dignity. Regulation can occur through policies adopted by the PTO and other agencies, court decisions, or, if necessary, legislation. Chapters 7, 8, and 9 will examine threats to science, medicine, and agriculture posed by DNA patents.

7

DNA Patents and Scientific Progress

INTRODUCTION

Chapters 5 and 6 considered some deontological arguments against DNA patents as well as some consequentialist arguments. The next three chapters will focus on consequentialist arguments against DNA patenting that address the impacts of DNA patenting on science, medicine, and agriculture. This chapter will consider the argument that DNA patents undermine scientific progress. Since consequentialist arguments ask us to consider possible (or probable) benefits and harms of actions and policies, this chapter will also make use of many of the points developed in chapter 6 for assessing consequences, including the application of the Precautionary Principle. Thus, this chapter will ask whether the threats to science posed by DNA patents are plausible; and, if so, what would be a reasonable response to those threats.

THE BENEFITS OF DNA PATENTS FOR SCIENCE, REVISITED

Before discussing the potential harms to science that may be brought about by DNA patents, it will be useful to review briefly the potential benefits, since benefit/harm decisions must balance benefits and harms. The potential benefits of DNA patents to science include:

- Incentives for researchers to conduct research with patentable applications;
- Incentives for companies and universities to invest in research that is likely to lead to patentable products and services;

- Increases in the pace of research due to competition among publicly and privately funded researchers for intellectual property and scientific discovery.

Thus, DNA patents, like other forms of intellectual property, can promote scientific progress by providing rewards and incentives for researchers and funding organizations. Although many scientists and scholars believe that knowledge should be its own reward, the practical realities of research do not conform to this intellectual ideal. Real world scientists usually have mixed motives: while they espouse the belief that knowledge should be pursued for its own sake, they are often motivated by such factors as career advancement, prestige, and the desire to acquire wealth (Resnik 1998b). Moreover, given the huge sums of money required to conduct research, research sponsors expect a decent return on their R & D investments. DNA patents supply rewards and incentives for individual researchers and sponsors in many disciplines that study DNA, including genetics, molecular biology, genomics, and proteomics.

How might one measure the impacts of DNA patents on progress? First, it will be useful to define "scientific progress." Progress is teleological notion, since to make progress is to take steps toward a goal or objective. *The American Heritage Dictionary* (2000) defines the verb progress as follows: "1. Movement, as toward a goal; advance. 2. Development or growth: students who show progress. 3. Steady improvement, as of a society or civilization: a believer in human progress." For example, someone who is making progress as a piano player is taking steps toward the goal of being a good piano player. A building contractor who is making progress in his construction of a house is getting closer to completion of the house.

While the concept of progress makes a great deal of sense when applied to the construction of a house or the development of a piano student, philosophers, sociologists, and historians of science have spent several decades arguing about the progress of science. The book that launched this thorny debate was Thomas Kuhn's *The Structure of Scientific Revolutions* (1962), which argued that scientists do not slowly accumulate knowledge in a steady march toward the truth. Science changes through revolutionary revisions of theories, methods, and assumptions known as paradigm shifts. In between revolutions, scientists accumulate data, solve problems, and develop theories and methods within a commonly accepted paradigm. Ever since the publication of Kuhn's book, scholars who study science have attempted to answer the following basic questions about scientific progress (Newton-Smith 1981):

1. Does science make progress?
2. Can we know that science make progress?
3. How can we measure scientific progress?

For the purposes of this book, I will assume that (1) science does make progress (in the long run) and (2) that we can know that science makes progress. Developing a detailed answer to these two assumptions would take us way beyond the scope of this book, so I will provide only a brief sketch of a view I will adopt for the purposes of the book. (See Popper 1959; Laudan 1978; Newton-Smith 1981; Kantorovich 1993; and Kitcher 1993 for further discussion.)

To understand scientific progress, we must first have an account of the goals of science, since progress is a teleological notion (Laudan 1978). Science is a complex human activity that cannot be defined by any single goal or purpose (Resnik 1993, 1999c); science fulfills a number of different goals, including:

- The acquisition of knowledge about the natural world;
- The explanation of natural phenomena;
- The development of laws, theories, and generalizations about the natural world;
- The prediction and control of nature;
- The improvement of human welfare;
- The solution of practical problems.

To get an accurate measure of the progress of science toward these different goals, one would need to compare the current state of development in a particular research field with its previous state of development to determine whether the field has gotten any closer to its goals (Niiniluoto 1999). By comparison, to measure the progress of a house, one can compare two successive states of the house with the end result that one is seeking (i.e., the completed house). Using the approach, one can could determine that one stage of the house, stage A, was closer to the goal than the previous stage, stage B, because, for example, stage A contained the foundation but no walls but stage B contained the foundation and walls. But this approach to measuring progress raises a difficult philosophical problem: how do we know what the end result of particular scientific discipline is supposed to be? How do we know what genetics will look like in some future time when the science is complete (or at least more complete than it is now)? With a house, we have blueprints, artistic renditions, and other ideas and documents that specify the end result. But we lack similar markers for scientific fields. We cannot gaze into a crystal ball and say where any particular field of research is or should be going. We can look back into the past with some confidence and assert that a particular scientific field has made progress, but we cannot compare the present state of any scientific field to some idealized future state (Newton-Smith 1981). It is therefore impossible to know or measure whether scientific fields make any progress toward anything, one might argue.

My strategy for dealing with this philosophical problem is to distinguish between direct and indirect measurements of progress. In a direct measurement of progress, one has a specification of the end result to use to compare two or more successive stages. We can make direct measurements of the progress of the construction of a house, but we cannot make such measurements on scientific fields. Nevertheless, we can measure progress indirectly by using surrogate end points. A surrogate end point is a result that tends to correlate with the sought after end point. Medical researchers frequently use surrogate end points in research on human health and disease. For example, in testing an HIV medication, such as zidovudine (AZT), a desired end point would be mortality; an effective medication would be one that reduces overall mortality from HIV. If one cannot obtain accurate mortality data on an HIV medication, one may still test its effectiveness by using surrogate end points that are associated with reduced mortality, such as reduction in viral load or increases in T cell counts. Researchers are justified in using the surrogate end points because they know that increases in viral load and decreases in T cell counts tend to be associated with increases in mortality from HIV infection (Jacobson et al, 1991).

What would be a useful surrogate end point for measuring scientific progress? To answer this question, one can look at the history of science to determine what types of phenomena are typically associated with progress. To conduct this survey, one must assume that there are some cases where we do know that a particular field has made progress, even if we do not have access to the desired end state of the field. For instance, I think most people would agree that computer science and genetics both made tremendous progress between 1953 and 2003. If these two fields do not constitute clear example of progress, then I do not know what would count as a good example of progress! What are some quantifiable characteristics of the fields during this period that we can use as surrogate end points for measuring progress? A short list might be as follows:

- The number of publications.
- The number of patents. (Since we are trying to determine how patents affect progress I will not use patents as a surrogate end point for measuring progress, since this would beg the question at issue. However, patents generally are useful surrogate end points that reflect the degree of technological development in a particular field that has practical applications. Although patents are not a useful surrogate end point for astrophysics, they are a useful surrogate end point in mechanical engineering.)
- The level of public and private funding.
- The number of people employed in the field.
- The degree of specialization (i.e., splitting into subfields, branching into new fields).

To understand how patenting has affected (or may affect) the progress of science, we need to ask ourselves whether the fields that study DNA have made more or less progress since the first DNA patents were awarded. Since we will not use patents as a surrogate end point, we can consider some of the other factors.

Let's consider two easy cases first. In 1976, the year the first DNA patent was awarded, the two main sciences that studied DNA were genetics and molecular biology. Since 1976, several new fields and subfields have emerged including genomics, proteomics, structural biology, plant genomics, comparative genomics, pharmacogenomics, and developmental genomics. As noted in chapter 4, the biotechnology industry has created thousands of jobs in the last two decades: many more people are employed in the genetic sciences today than were employed in these fields in 1976. Based on these two measures, the increase in specialization and the increase in employment, the genetic sciences have made considerable progress since 1976.

In response to this evidence, one might argue that increases in specialization and employment do not always correlate with progress. For example, from the late 1800s to the early 1900s, many scientists latched onto the N ray bandwagon and were employed as N ray scientists. N rays were supposed to be invisible forms of radiation similar to X rays or radio waves. Between 1900 and 1903, many scientists joined this new field, and N ray researchers published over 100 papers on the topic. As it turned out, N rays did not exist. They were nothing more than a large scale case of self-deception based on an observer effect. People observed N rays because they expected to observe them (Broad and Wade 1993). My response to this objection is that the genetic sciences are not like N rays. Geneticists study real phenomena not illusions reinforced by active imaginations.

With regard to funding, chapter 1 mentioned that private funding for biomedical R & D rose from $2 billion per year in 1980 to over $36 billion per year in 2000 (Resnik 2001a). Public funding for biomedical R & D has also risen dramatically since 1980. The budget of NIH, which provides most of the public funding for biomedical R & D in the United States, has risen dramatically in the last five years. The NIH budget rose from $5.5 billion in 1985, to $10.9 billion in 1995, to $15.6 billion in 1999, to $20 billion in 2001, to $23 billion in 2002, and to $27 billion in 2003 (NIH 2001, 2002). Although it is not at all clear that patents have played an important role in the dramatic increases of public funding, there can be little doubt that they have played a key role in the huge increases in private funding of biomedical R & D, since private corporations do not pour billions of dollars in R & D without expecting reasonable returns on their investments. In response to this evidence, an opponent of DNA patents could argue that increases in funding do not correlate with scientific progress. Various governments and private corporations have also wasted money on dubious causes, such as polywater, laetrile, N rays, and cold fusion. I am not thoroughly convinced that increases in funding are good measures of progress although I think such evidence needs to be taken seriously and should not be ignored.

The publication record, on the other hand, provides a much better indicator of progress. Although the publication record is not a perfect surrogate end point for progress—remember the N ray episode—it is perhaps the best surrogate end point that we have. It is also fairly easy to quantify. To provide some measure of scientific progress in the genetic sciences (i.e., the fields of genetics, molecular biology, and genomics), I obtained data from the Science Citation Index and Medline, two of the premiere scientific research indexing and abstracting services (see table 7.1). I searched these databases using a Boolean search with the term "gene or gene or DNA or genetics or genomics" for the period 1991–2002. Although these databases go back earlier than 1991, I did not include these data due to reporting inaccuracies. Both of the databases were (at the time) still completing their backlogs for the years before 1991, and the data were incomplete. Not surprisingly, I found that publications in the genetics sciences increased streadily from 1991–2002: the Science Citation Index showed an increase from 43,089 to 114,354, and Medline showed an increase from 54,003 to 113,442. So the publication rate has more than doubled by both of these measurements.

To determine whether patents have helped or hindered progress, I obtained data from the PTO's patent database using "DNA" as a search term in the field of "claims" (see table 7.2). This search provided a precise measurement of the patents issued by the United States that make claims pertaining to DNA. The results of this search indicate that the first DNA patents were awarded in 1976 and that the number of patents has risen from fewer than 100 per year prior to 1987 to 2000 or more per year by 2002. The PTO has awarded an estimated 16,380 DNA patents since 1976. Thus, these data indicate that patents and publications in the genetic science both increased from 1991–2002.

To better understand the relationship between patents and publications, I correlated patents with data from the Science Citation Index (SCI) and from Medline. I found that the correlation coefficient (a measure of correlation) between patents and publications in the SCI from 1991–2002 was 0.895, indicating a very strong positive correlation. This number also had a statistical significance (or p-value) of 0.01, which indicates that there is only a 1 percent chance that this observed correlation was due to random variations in the data. I also found a very strong positive correlation between patents and publications in Medline during the period. The correlation for these two variables was 0.931, and the p value was 0.01.

Patent activity correlates positively with the publication rate: as patents have increased, so have publications (see figure 7.1). This valuable piece of information does not prove a cause and effect relationship between patents and publication, however, because we do not know the order of causation (do

TABLE 7.1
DNA Patents and Publications in the Genetic Sciences, 1991–2002

Year	DNA Patents	SCI	Medline
1991	357	43,089	54,003
1992	431	48,397	59,264
1993	569	54,714	64,117
1994	544	60,301	70,675
1995	603	67,307	76,906
1996	1,006	71,566	80,643
1997	1,496	76,906	84,633
1998	2,078	82,571	91,642
1999	2,066	85,461	97,659
2000	1,896	87,731	106,064
2001	2,143	92,995	110,067
2002 (projected)	2,018	114,354	113,442
Totals	15,217	885,392	1,009,096

Correlation between Patents and SCI: $r = 0.895$; p-value = 0.01
Correlation between Patents and Medline: $r = 0.931$; p-value = 0.01

These data were obtained by searching the Science Citation Index (SCI) and Medline, using the Boolean search "gene or genome or DNA or genomics or genetics," for the years 1991–2002. Search date: August 20, 2002.

Although these databases contain data prior to 1991, these data are not included due to inaccurate reporting because the databases were incomplete for these earlier years.

patents cause publication or vice versa?) and the correlation could be spurious (i.e., a third variable might explain the correlations we observe). For example, increases in public and private funding during the period could explain why we observe this correlation, because funding leads to publication and to patents. On the other hand, these strong correlations do provide evidence that patents are not impeding publication. If patents were impeding publication, we would expect to see a negative relationship between patents and publications. In any case, those who argue that DNA patents are harming science will have to find a way to account for these strong correlations between patents and publication because they provide convincing evidence that DNA patents are helping rather than hindering research in the genetic sciences.

Opponents of DNA patents could argue that researchers would be publishing at an even greater rate if patents were not available (Campbell 2002). Although this hypothesis is possible, it would seem to be very difficult to obtain evidence that would support such a counterfactual claim. Perhaps the better reply would be to assert that the observed rate of progress will one day

TABLE 7.2
DNA Patents Issues by the U.S. Patent and Trademark Office, 1976–2002

Year	Number of DNA Patents Issued
2002 (projected)	2,018
2001	2,143
2000	1,896
1999	2,066
1998	2,078
1997	1,496
1996	1,006
1995	603
1994	554
1993	569
1992	431
1991	357
1990	265
1989	253
1988	179
1987	144
1986	70
1985	57
1984	53
1983	36
1982	36
1981	18
1980	10
1979	6
1978	7
1977	4
1976	2
Totals	16,380

Data obtained by searching the U.S. PTO's database using "DNA" as a search term in the search field "claims." These data represent all patents awarded that make claims pertaining to DNA.

slow down, due to the adverse effects of patenting. The short-term gains associated with patenting will eventually lead to greater losses in the long run. Thus, we should consider how patenting is likely to affect that future of research. Opponents of DNA patents see a grim forecast for that future. How might some of these bad results arise? In the remainder of this chapter I will consider and evaluate some of the arguments for these prognostications.

DNA Patents

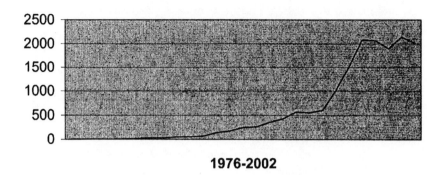

1976-2002

Publications in the Genetic Sciences

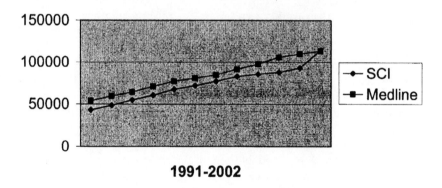

1991-2002

FIGURE 7.1. DNA Patents and Publications in the Genetic Sciences

SECRECY

The ethic of openness has played a crucial role in the incredible progress of science that has occurred in the last 600 years. Openness contributes to the progress of science in three ways. First, by sharing ideas, methods, and information researchers cooperate and work together on common problems, which increases the amount of resources devoted to those problems (Munthe and Welin 1996). Many hands lighten the load. Second, by sharing ideas, methods, and information researchers are able to scrutinize and critique each other's work. Openness plays a key role in peer review and replication. Third, by sharing ideas, methods, and information researchers are able to develop new insights and hypotheses. Openness promotes cross-fertilization of ideas that leads to creativity (Resnik 1998b).

It may come as a surprise to many readers to learn that scientists did not always share ideas freely. During the 1300s and 1400s, many researchers kept their ideas a secret because they were afraid of having them stolen. This attitude began to change with the advent of the patent system and the printing press in the 1500s, and the publication of the first scientific journals in the 1600s and 1700s, as scientists began to value publication and public dissemination. However, scientists still demanded proper credit for their accomplishments, as well as proper citation, even though they were willing to publish their ideas. Moreover, scientists kept their research secret prior to publication in order to preserve their interests in priority and protect their reputations and integrity. For example, Charles Darwin waited more than twenty years to publish his theory of evolution by natural selection in order to obtain evidence for the theory and organize his thoughts.

Today, the conflict between secrecy and openness in science continues. Although most researchers are strongly motivated to share information, they sometimes keep their work a secret in order to (1) protect their interests in priority, (2) preserve their reputations, (3) prevent outside parties from interfering in their work, and (4) to protect trade secrets and private research, confidential information pertaining to human subjects, publications submitted for peer review, and classified military research (Resnik 1998b). However, an increasingly important reason why researchers sometimes keep research under wraps is to protect potential patents rights and other intellectual property claims.

When research may have patentable applications, data and other research records play a crucial role in establishing the originality of the invention. If an invention has been previously disclosed in the prior art by means of publication, patenting, or use, then it cannot be patented, according to U.S. patent law (see chapter 3). In order to settle disputes related to novelty when two parties have submitted patent applications covering the same claims, the PTO will consider evidence that demonstrates who was the first to conceive of the invention. As noted in chapter 3, the first person to conceive of the invention will be able to patent the invention, provided he exhibits due diligence in reducing his invention to practice and submitting a patent application. If an inventor has failed to exhibit due diligence, then the PTO will attempt to determine who reduced the invention to practice first. All of these inquiries require the PTO to have access to data and records. Since an inventor who discloses information related to his invention prior to patenting may lose his patents rights, there is considerable pressure to keep research records confidential prior to filing a patent claim. An inadvertent disclosure to the wrong

person can also undermine one's hopes of patenting an invention, if that outside party steals one's idea and submits a patent application. Once a claim has
been filed, researchers may publish and share their work without the fear that
they will lose their patent rights.

There is a growing body of evidence indicating that researchers may delay
or forego publication in order to protect intellectual property claims. A survey
conducted by Blumenthal et al. (1997) of 3,394 life science faculty found that
19.8 percent had delayed publication by at least six months in the last three
years to allow for a patent application, to protect a scientific lead, to allow time
to negotiate a patent, or to resolve intellectual property disputes. The survey
also found that withholding was most common among the most productive
faculty. Campbell et al. (2002) obtained similar results in a survey of 1,849
geneticists. Forty-seven percent of the researchers in this survey reported that
they had been denied at least one request for data, information, or materials
in the last three years. Twenty-eight percent of geneticists said they were
unable to confirm published research due to a lack of access to data. Thirty-
five percent of those surveyed said that sharing had decreased in the last
decade, although 14 percent said that sharing had increased.

Very often, private companies require their researchers to delay or with-
hold publication in order to protect patents. Eighty-two percent of biotechnology companies sometimes require scientists to not disclose results prior to
the filing of a patent, according to one survey (Gibbs 1996). Researchers often
sign contracts with private companies that forbid them from sharing data,
tools, methods, or resources without the company's express permission.
Indeed, one survey found that 53 percent of scientists at Carnegie Mellon
University had signed such contracts allowing companies to delay publication
(Gibbs 1996). Finally, a survey of 1,105 authors of articles published in biomedical journals showed that 10 percent of those articles contained disclosures about current or pending patents (Krimsky et al. 1996). Since this survey took place before the renewed awareness about conflicts of interest in
research, 10 percent is probably a low estimate.

Clearly, since openness is such an important part of the scientific ethos,
researchers and concerned citizens should be wary of threats to openness
posed by intellectual property concerns, such as potential patents and commercial interests (Knoppers, Hirtle, and Glass 1999). But what is the precise
nature of the threat to openness posed by DNA patents? Is it a plausible
threat? Is it a uniquely dangerous threat? What would be a reasonable
response to this threat? Readers will recall the notion of a plausible threat
discussed in the previous chapter. To be plausible a threat must be more than
a fanciful tale; it must have some evidence in its favor. The evidence presented thus far indicates that the threats to openness are indeed plausible.
But how dangerous are those threats? Is the danger so great that it cannot be
managed or mitigated? There are three reasons for believing that the threats

to openness posed by DNA patents, though plausible and important, can nevertheless be controlled and mitigated.

The first reason for thinking that DNA patents do not pose a dangerous threat to openness is that the period of secrecy prior to the filing of a DNA patent application will end once the patent is awarded and the application enters the public domain. For example, suppose it takes a biotechnology company seven years to develop a genetic test for breast cancer mutations. During that period of time, it keeps its data secret and does not publish the data. If it takes the patent agency two years to approve the patent, then the data will be kept secret for nine years before entering the public domain. Once the researcher obtains a patent, he or she can also publish data contained in the patent application, even though the data will be available through the patent agency. Although nine years is a significant period of time, it is not a very long time when considered against the background of the history of science. For example, Darwin kept his work a secret for over two decades before publishing his *Origin of Species,* but this period of secrecy certainly did not hinder the study of evolution. Watson and Crick showed only a few people their model of the structure of DNA prior to winning the Nobel Prize in 1953. Mendel's work on the laws of inheritance was ignored for several decades before it was rediscovered at the beginning of the twentieth century (Meadows 1992). The important point here is that secrets that lead to patentable inventions are eventually revealed. A reasonable period of secrecy prior to public dissemination does not do irreparable harm to science and may even help researchers protect research from premature disclosure. A far greater threat to science is posed by trade secrets, which can remain secret as long as a company can maintain secrecy. But, as we have already seen, patents discourage trade secrecy and encourage public disclosure.

Some changes in patent policy, if adopted by the United States and other countries, could help ease some of the pressures to maintain secrecy prior to submission of a patent application. In the United States and other countries, prior publication of an invention undermines the patent applicant's novelty claims. Since researchers do not want to risk having their ideas stolen or the patents invalidated, they usually wait until they obtain patent rights before publishing any data related to an invention. Patent agencies could relieve some of these pressures to maintain confidentiality by using more extensive use of provisional patent applications. The United States has allowed inventors to file provisional patent applications since 1995. A provisional application is not a formal patent application but it notifies the agency of one's intent to file a formal application (U.S. Patent Act 1995). In the United States, a provisional patent lasts for one year, after which time an inventor must file a new, formal application or convert the provisional application into a formal application. If an inventor chooses the latter option, the patent term begins running at the beginning of the submission of the provisional application. Inventors filing provisional applications may write the phrase "patent pending" on publicity related to their inventions.

Countries could use provisional patents to ease some of the secrecy pressures. They could extend the life of provisional patents beyond the one-year limit. Three years might be a more appropriate period. They could give inventors explicit permission to publish data related to their inventions during the provisional period, provided that inventors disclose their publication plans to the agency prior to publication to ensure that publication of data will not constitute a disclosure that invalidates the patent.

The rationale for this proposed change in policy is that the requirement of novelty should be used to show that the inventor was the first to conceive of the invention. If this is the goal of the requirement, then it should not matter whether the inventors have already published data related to the invention, as long as they are the first to publish. Their own publications will provide additional evidence that they were the first to conceive of the invention; publications should not invalidate the patent. Thus, if a researcher contributes DNA sequence data to a public genetic database, this prior act of publication should not prohibit the researcher from obtaining a patent. A policy such as this would help to reduce the period of secrecy in patentable research and could encourage more publication.

The second reason for thinking that DNA patents do not pose a uniquely dangerous threat to openness is that biomedical researchers and biotechnology companies have many other intellectual property interests besides DNA patents. If researchers were unable to patent DNA, they would still try to patent reagents, proteins, hormones, cell lines, tissues, genetically engineered plants and animals, and so on. Eliminating DNA patents would not stop or radically curtail secrecy in research related to intellectual property in biomedicine. The only way to eliminate the threat to openness posed by patents in biomedicine would be to forbid the patenting of all biological materials, including biochemically significant molecules. This would not be a realistic or workable option. Indeed, a ban on DNA patents would only encourage the patenting of downstream products of DNA, such as RNA and proteins.

A third and final reason for thinking that DNA patents do not pose a uniquely dangerous threat to openness is that intellectual property concerns are probably not the main reason why researchers in genetics and other biomedical sciences do not share data. According to the survey conducted by Campbell et al. (2002), among those geneticists who said they had withheld postpublication of data, information or materials, eighty percent cited the effort required to produce materials or information as a reason to not share; sixty-four percent cited need to protect publication interests; forty-five percent cited the financial cost of providing materials of information; twenty-seven percent cited obligations to industrial sponsors; twenty-four percent cited duties to protect patient confidentiality, and only twenty-one percent cited the desire to protect the commercial value of their results. Thus, researchers in the biomedical sciences would still have many different reasons not to share data, even if patents on biotechnology were not available.

CONFLICT OF INTEREST

In the last decade or so there has been a growing concern about conflicts of interest in research as a result of the financial interests of researchers and research institutions. A conflict of interest (COI) for an individual researcher can be defined as a situation in which the researcher has a personal or financial interest that undermines the ability to perform primary ethical or legal obligations in research. A conflict of interest for an institution can be defined as a situation in which the institution has financial or political interests that undermine its ability to carry out its primary ethical or legal obligations in research (Resnik and Shamoo 2002). Researchers and institutions have a variety of ethical obligations, including duties pertaining to honesty, objectivity, and openness; duties to give proper credit and respect intellectual property; duties to protect human and animal subjects, and duties educate and mentor students, to name but a few (Resnik 1998b). COIs undermine research by increasing the risks of biases at various stages of research, including problem selection, funding, subject recruitment, testing, data analysis, data interpretation, data management, and dissemination of results (Resnik 2000). Although COIs are not the same thing as research misconduct, they increase the risks of various types of misconduct, such as fabrication and falsification of data (Bodenheimer 2000). Even when COIs do not affect the quality or integrity of research results, they can undermine trust among researchers and between researchers and the public (DeAngelis 2000). Researchers should also be concerned about apparent COIs (i.e., COIs that cause the perception of bias even when they do not actually lead to bias).

In the last decade or so, researchers have obtained empirical evidence concerning COIs in research. A growing body of evidence indicates that many researchers, especially researchers in the biomedical sciences, have significant financial interests related to their work, including stock or stock options, consulting fees, honoraria, and patents rights (Blumenthal et al. 1996; Krimsky et al. 1996; Korn 2000). Other studies demonstrate some of the effects of COIs on research. Especially striking is the strong correlation between the source of funding and research outcomes in medical research. If a pharmaceutical company funds a study, it is very likely (a 90 percent chance or more in some cases) that the study will favor that company's product (Davidson 1986; Hess 1999; Yaphe et al. 2001).

In order to deal with the problem of COIs in research, institutions, government agencies, journals, and professional associations have developed several different strategies including disclosure, management, and prohibition (Korn 2000). When one discloses a COI, one informs a relevant, independent party, such as an employer or granting agency, about one's financial or personal interests. In order to manage a COI, those with the COI and their supervisor(s) must develop formal plans and procedures for minimizing the biases and other problems related to the COI. When an institution or organization prohibits a COI, it adopts rules or policies that forbid situations that create specific types of COIs or require people to avoid those situations. In the political arena, ethics rules require politicians to disclose significant gifts and political contributions and they forbid politicians from accepting money for votes. Research indicates that while almost all universities have COI policies, they also exhibit a great deal of variation concerning the content of those policies (McCrary et al. 2000). Almost all universities require disclosure, but few universities prohibit COIs. Although most COI policies adopted by institutions tend to focus on controlling individual COIs, there is growing recognition of the need to respond to institutional COIs as well (Public Health Services 2000; Resnik and Shamoo 2002).

Patents can create COIs and apparent COIs for researchers and researcher institutions, because a researcher with a patent (or pending patent) may become biased as a result of financial motivations. For instance, a researcher might consciously (or subconsciously) skew data or overestimate its significance. A researcher might also pressure subjects to enroll in clinical trials in order to collect data that could support his invention (see chapter 8). To handle these potential COIs, most COI policies adopted by journals, granting agencies, and research institutions require researchers to disclose patents or pending patents. DNA patents clearly pose a plausible threat to the objectivity and trustworthiness of biomedical research. But is that threat so uniquely dangerous that it cannot be adequately controlled or mitigated? Recent policy developments related to COIs in research indicate that researchers, government agencies, professional associations, journals, universities, and private companies have the resources and the wherewithal to deal with COIs in research, including those related to DNA patents (Morin et al. 2002). It will take some time to develop and implement effective policies, but the biomedical research community is acutely aware of the problem and what needs to be done about it. One would expect that in the future institutions will continue to develop, refine, and implement COI policies, and that these policies will deal with COIs related to patenting.

Moreover, the COI problem is not unique to DNA patents, since researchers, companies, and universities have other intellectual property interests, as well as other financial interests in research. The COI problems created by DNA patents are miniscule when compared to COI problems relating to pharmaceutical patents or the COI problems relating to the ownership of stock and other forms of remuneration. Banning DNA patents would therefore have only a minor impact on financial interests in research, since researchers would still have many other financial interests that would need to be managed. The most practical solution is to develop and implement effective policies for dealing with all of the various types of COIs in research (Morin 2002). Some key points for COI policies relating to patents would include:

- Disclosure of patents or pending patents on papers submitted for publication, grants submitted for review, and papers presented at meetings.
- Development of management plans for individual researchers who create their own start up companies (i.e., companies that are established to develop and test their inventions).
- More discussion of how to effectively deal with institutional COIs, especially financial ties between universities and corporations.

We will discuss COIs again in chapter 8, when we look specifically at their effects on medicine.

BIOTECHNOLOGY AND THE FREE MARKET

As discussed in chapter 3, patent holders have the right to license others to make, use, or commercialize a patented invention. Patent holders are not required make, use, or commercialize their inventions or license anyone to do so. Several commentators have argued that problems related to free market dynamics could harm progress in biotechnology by blocking the development of new technologies or by creating upstream tolls on downstream research (Heller and Eisenberg 1998; Merz et al. 1997; Merz and Cho 1998; Merz 2000; Guenin 1996; Svatos 1996; Andre 1992; Andrews and Nelkin 2000; Henry et al. 2002). In chapter 3 we also noted that the main rationale for the patent system is to provide incentives and rewards for inventors and entrepreneurs. The rationale assumes that a free market approach is the most effective way of developing, utilizing, and distributing information and technology. However, markets do not always operate efficiently, and a variety of problems related to market inefficiencies and distortions can arise. According to some commentators, adverse effects of the free market can result in a genetic anti-commons in which genetic information is underutilized and new products are not developed (Heller and Eisenberg 1998). This section will examine some of these potential problems and suggest some ways of avoiding or mitigating these difficulties.

In any given domain of industry, different individuals, companies, or organizations may hold patents and copyrights on different products, processes, techniques, and methods. A great deal of useful information may also exist in the public domain. For example, in the computer industry, a great deal of information pertaining to mathematics and logic is in the public domain, but different companies own patents on different types of hardware and software. In order to develop a new product in the computer industry, a company may need to obtain licenses from other companies in order to use their inventions and avoid patent or copyright infringement. As we noted in chapter 3, one may infringe a patent on a part by using that part in another invention without permission of the patent holder (*Eastman Kodak Co. v Image Technical Services, Inc.* 1992). For example, if company X has a patent on a particular integrated circuit and company Y is developing a computer that uses that circuit, then the two companies may negotiate a license allowing company Y to use the circuit in its new product, in exchange for royalties. Of course, company X could also refuse to grant any company a license, since the United States has no compulsory licensing law. The same situation would arise if a company is planning to use software in developing a new product. In academia, licenses become a major concern when a professor decides to reprint previously published articles in a textbook. In order to publish the textbook, permissions must be obtained from the copyright holders.

In both industry and academia, we can the distinction between upstream versus downstream research to elucidate licensing issues. Upstream research consists of basic technologies (e.g., products, tools, and ideas) that are applied by downstream users to develop new technologies. In computer science, transistors are upstream inventions and handheld computers are downstream inventions. Systems of logic and mathematical algorithms are upstream ideas and computer programs are downstream applications. In order to develop new products, downstream researchers may use information in the public domain, but they may also need to negotiate licenses with upstream researchers.

Since companies may often need to use thousands of different patented inventions or thousands of copyrighted original works, they often sign general licensing agreements with one another, in order to minimize the administrative and legal costs associated with negotiating licenses on a case-by-case basis. Although it is not always easy to negotiate licensing agreements, and companies may fail to reach agreements from time to time, corporations are familiar with these legal difficulties and they are often strongly motivated to cooperate when it comes to licenses because they recognize their interdependence and the high costs of infringement lawsuits (Scott 2000; Woollett and Hammond 1999). A company that refuses to license another company to use its products may need that other company's products at some point, and a company that is not careful to negotiate licenses may have to defend itself against lawsuits for direct or indirect infringement.

Licensing problems can also occur in the pharmaceutical and biotechnology industries. A great deal of useful information in these industries has been placed in the public domain through the publication of journal articles, databases, or patent applications. However, a great deal of information is also patented or copyrighted. Consider the potential licensing nightmare for developing a genetic test for mutations related to cystic fibrosis (CF). There are over six hundred mutations related to CF, which are currently not patented. The entire human genome is now in the public domain, so a company developing a CF test will be able to use this information. Suppose that a dozen different companies patent different CF mutations and that their patents give them control over the use of these mutations to test for CF. In order to develop an assay that tests for all six hundred plus mutations, a company might need to negotiate a dozen different licenses. (If one company patents all of the mutations, this problem does not arise.) A similar sort of problem could arise for someone attempting to develop a gene therapy treatment for CF. To develop and market this gene therapy product, a company might need to obtain permission to use patented DNA sequences, proteins, vectors, methods, and procedures. The same sorts of licensing problems could arise in virtually any area of technological development where there are patents on DNA, including agricultural biotechnology. The reason why these problems occur is that one cannot use a patented invention to make, use, or market a new invention, without permission of the patent holder. If X designs a new computer with 500 different patented components, then he will need to negotiate licenses on all 500 of these inventions in order to make, use, and market his computer.

Given the possibility that many different companies and individuals may have IP rights pertaining to different products and processes in biotechnology, it is likely that companies will have difficulty negotiating these different licenses. To develop a new genetic test or treatment, a company might have to negotiate dozens of different licenses and pay a variety of licensing fees. In addition, a company may need to pay fees to have access to genetic databases and information services developed by companies. If the web of licensing agreements and various fees becomes too complex, costly, and burdensome some researchers and companies may decide not to develop new technologies since every licensing agreement will incur paperwork, legal fees, licensing fees, user fees, and so on.

There is also no guarantee that all parties will cooperate or negotiate in good faith. Some companies may refuse to grant licenses in order to maintain a competitive advantage in the market. A company may develop or acquire a patent in order to block or prevent another company from developing a competing product. As noted in chapter 3, this strategy is legal in the United States, since the United States does not have compulsory licensing. However, one might argue that compulsory licensing is unethical because it defeats the purposes of a patent system designed to encourage product development (Guenin 1996).

The net result of these potential licensing problems is that DNA patents could pose enormous problems for downstream research and applications. These tolls could increase the cost of developing new products and conducting new research. Companies may pass on these costs to consumers. Some companies may decide not to develop a particular product because they do not want to deal with licensing hassles. While private corporations are accustomed to dealing with licensing problems, academic researchers often lack the legal or financial resources to do so. Thus, licensing problems could hinder commercial research and have a chilling effect on academic research. Indeed, one survey reveals that researchers are very concerned about these licensing issues (Schissell, Merz, and Cho 1999). The net result of this problem, according some scholars, is that new technologies will be blocked from the market and genetic information will be underutilized (Heller and Eisenberg 1998). A genetic anticommons could emerge.[1] Instead of being an efficient developer and distributor of genetic technology, the free market might have the opposite effect. One recent case provides us with some anecdotal evidence of problems associated with licenses in biotechnology. As many as seventy different intellectual property claims have had some impact on the gene transfer techniques needed to produce genetically modified (GM) rice enriched with beta-carotene. After having licensing difficulties for several years, Monsanto Corporation agreed to provide cost-free licenses to help accelerate the development of GM rice (Normile 2000).

To work around these licensing problems, some companies may develop copycat inventions to avoid licensing costs or to compete with patented inventions. A copycat invention is one that barely meets the novelty test for patentability. For instance, suppose one company has patented a very successful antidepressant medication that it refuses to license other companies in order to maintain its monopoly. Another company might develop a medication functionally identical to the existing medication with a few structural differences. The new medication will pass the novelty test if it is judged to be a slight improvement on the existing medication. Copycat inventions that are improvements on existing inventions benefit science, industry, and the public. Trivial copycats, on the other hand, may be regarded as a waste of R & D resources and the consumer's money. Trivial copycats are fairly common in the pharmaceutical industry, where companies may develop copycat drugs in attempt to try to capture some of the market for a blockbuster drug, such as Prozac (Stolberg and Gerth 2000).

Advocates for DNA patents argue that the free market will take care of these licensing issues (Tribble 1998; Wollett and Hammond 1999; Scott 2000). Although licensing difficulties may hold up some technologies while negotiations take place, in the long run, various parties will reach agreements because they will realize that it is in their economic interests to cooperate and negotiate. As long as different companies control different technologies and no company has exclusive control over a significant portion of the market, companies that want to develop new products will not be able to do so without negotiating with other companies. Companies will cooperate not because they are altruistically motivated, but because they recognize the mutual advantages of cooperation. Two major players in the biotechnology, Celera and Incyte Pharmaceuticals, have stated that they plan to make most of their money from selling information services and from licensing (Marshall 1999a).

Blocking patents will probably play a minor role in biotechnology because they will not be profitable. Most biotechnology companies have business plans that focus on product development and marketing, licensing, or information services. Blocking patents are most effective when a company controls a significant portion of the technology in a particular area and can profit from keeping competing products off the market. The costs of licensing fees will also not pose a significant barrier to research and product development because the market will drive down excessively high fees due to decreased demand. These fees may inflate the costs of doing research in biotechnology, of course, but someone must bear the cost of research. Licensing fees are one way private companies can attain a return on their investments. As long as no single company has exclusive control over a significant portion of the market, the business climate will encourage self-interested cooperation.

Copycat inventions will probably not cause any great harm to the industry or society. Not every invention needs to be a major advance or improvement in order to benefit society. Minor improvements can also contribute to the progress and science and technology and provide added competition on the market, even if they make less of a contribution than highly innovative inventions. Many companies will refrain from producing copycat inventions because the copycats will not be profitable. Moreover, patent agencies have had to deal with the problem of copycat inventions for years, and they have managed to develop policies that allow for minor improvements but control the problem of trivial copycats.

Recent developments provide evidence of cooperation in the biotechnology industry. For the most part, companies have shown a willingness to work together in negotiating licensing agreements (Marshall 1999a). Recognizing the need for access to DNA sequence data, private companies, universities, and government organizations have pooled their resources to develop public DNA databases and consortiums, such a Genbank. As noted in chapter 2, public and private researchers have now made the entire human genome available through free and open databases. Celera has also published the genome of the fruit fly *Drosophila melanogaster,* and plans to make other genomes available to the public (Adams et al. 2000). Monsanto has decided to make the rice genome available to the public as well. Over a dozen companies, universities, and government agencies have developed a consortium for single nucleotide polymorphisms (SNPs) (Roberts 2000). Human Genome Sciences, which recently patented a human gene that allows HIV to enter the cell, plans to share data and reagents with researchers. Like Celera, the company will allow free access to genetic information for research purposes, but it will charge licensing fees for marketable products developed from that information (Marshall 2000a,c).

However, one might argue that this spirit of cooperation could vanish if biotechnology and pharmaceutical companies face increasing financial pressures (Heller and Eisenberg 1998). Although licensing problems are not currently a significant problem in biotechnology, this could quickly change if companies no longer find it profitable to cooperate. There is some evidence that the spirit of cooperation in existence today is not entirely harmonious or stable. For example, negotiations concerning public and private collaboration on sequencing the human genome have not always gone smoothly, and there have been disputes about data sharing and access (Marshall 2000b,c).

Furthermore, the emergence of dominant companies could undermine the efficiency of the free market in this area. If a single company were able to acquire patents on many different DNA sequences, proteins, and other technologies, it might be able to develop a monopoly in an area of biotechnology. Or perhaps a company might attempt to develop a monopoly based on a key patent, such as a patent on PCR. If such a dominant company emerged, it would have little incentive to cooperate with other parties, since it would have cornered the market in biotechnology. Even if no giant company emerges, a similar scenario could arise if several large corporations drove all the other corporations out of business and conspired to take steps to prevent other companies from gaining access to information and technology. Thus, the biotechnology industry, like other industries, must still face potential problems related to unfair competition and monopolies.

To summarize, the problems and concerns discussed in this section pose credible threats to the progress of science and should be taken seriously. Do these problems create such a huge risk that we should ban DNA patents? I think banning would be an extreme measure that is not called for at this time. Most of these licensing problems would exist with or without DNA patents. All of the same issues arise concerning proteins, cell lines, GM animals and plants, methods, procedures and other patentable inventions in biotechnology. DNA patents pose additional licensing problems for scientists and industry, but they do not pose any unique or especially dangerous threats. It is important that we do not forget that a great deal of data is constantly entering the public domain. Researchers have access to genetic databases, protein databases, patent applications, and many other important sources of information. The percentage of scientific and technical information under private control has been and probably always relatively small. Despite fears about a growing anticommons, there probably always will be a very large commons in biotechnology.

On the other hand, the problems related to free market dynamics call for some form of precautionary action designed to encourage publication of data and to counteract dominant companies. The following policies, if implemented effectively, could help overcome some of these problems relating to market inefficiencies in biotechnology:

1. Patenting agencies and the courts should pay careful attention to the requirements for obtaining DNA patents, making sure to achieve the proper balance of private control versus public access. Thus, they should make sure the scope of a DNA patent is broad enough to provide inventors and investors with incentives, but not so broad as to prevent competitors from entering the market. Courts should not allow inventors to abuse their patent rights by attempting to extend the scope of a patent beyond its original bounds. They should reject DNA patent applications that fail to state a definite and plausible utility. They should adjust the obviousness requirement so that it rewards researchers for insight and hard work, not for trivial modifications to the prior art.

2. Patenting agencies and the courts should develop, clarify, and use the research exemption to allow researchers to use patented DNA inventions for conducting research for non-commercial purposes (Nuffield Council on Bioethics 2002). This exemption should not extend to researchers conducting research for commercial purposes. Since the line between academic and nonacademic research has blurred in the last two decades, the legal system will have to carefully define this research exception so that it is neither too narrow nor too broad (Karp 1991). If the exemption is too narrow, then too few people will qualify for the exemption, and it will not benefit research. On the other hand, if too many people qualify for the exemption, it can undermine commercial intellectual property interests.

3. Federal agencies and the courts should also be mindful of unfair patenting practices and antitrust concerns. If need be, antitrust law can be applied to patents in order to prevent monopolies and unfair competition (*United States v. White Motor Co.* 1961).

SUMMARY

This chapter has identified some potential threats to the progress of science and technology posed by DNA patents, including increased secrecy, conflicts of interest, and problems related to free market dynamics. It has argued that while these threats to scientific and technological progress are credible, they do not pose such a great danger that we should ban DNA patents. The most prudent response is to take precautionary measures designed to both encourage investment in biotechnology and promote scientific progress. Society should adopt polices, discussed in this chapter, that strike a balance between public and private control of DNA. These policies include making more extensive use of provisional patents, developing strategies to manage conflicts of interests, and adopting policies designed to overcome market inefficiencies.

8

DNA Patents and Medicine

INTRODUCTION

Chapter 7 examined potential benefits and harms to science resulting from DNA patents. This chapter continues this type of benefit-harm analysis in the medical realm. After reviewing the benefits of DNA patents for medicine, the chapter will examine the potential harms. This chapter, like the previous two chapters, will apply the Precautionary Principle to the issues at hand. It will examine potential harms to determine whether they represent plausible threats, and, if so, what would constitute a reasonable response to them.

THE BENEFITS OF DNA PATENTS FOR MEDICINE, REVISITED

Before discussing the potential harms to medicine that may be brought about by DNA patents, it will be useful to briefly review the potential benefits, since benefit/harm decisions must balance both benefits and harms. The potential benefits of DNA patents are all related to the effect that patents have on bio-medical research, since research is essential to improvements in therapy (i.e., diagnosis, treatment, and prevention). Research includes basic research in fields such as genetics, genomics, proteomics, biochemistry, and microbiology, as well clinical and applied research in fields such as gene therapy, hematology, oncology, immunology, and endocrinology. Because researchers have been able to patent DNA only since 1976, and the biotechnology revolution began in the 1980s, most of the clinical and practical impacts of DNA patents will be realized in the future.

Without a doubt, patents have played a major role in the development of clinical applications of genetics. Although the fruits of genetic medicine may not arrive for many years, we have already seen a substantial number of patents on gene therapy techniques, genetic tests, as well as drugs derived from genetic manipulation, such as Epogen and human clotting factors (Fons 2000; Pabst 1999; Baggot 1998; Leonard 1999). Although ethical and policy discussions have focused, for the most part, on patents on genetic tests, there have been far more patents on gene therapy procedures, methods, tools, and products (Caulfield and Gold 2000a,b). Many different companies are now sponsoring gene therapy clinical trials on diseases such as coronary artery disease, cancer, liver disease, hemophilia, and autoimmune diseases (Gura 2001).

The Genetic Alliance, which represents over 300 patient advocacy groups for people with genetic diseases, supports DNA patents as a means of providing incentives for private companies to sponsor basic research and conduct clinical trials (Genetic Alliance 2000). Although the organization is not proindustry, it recognizes the importance of private investment and involvement in transferring technology from the laboratory to the clinical setting. It usually takes more than a decade as well as hundreds of millions of dollars to develop a new drug or medical device. In order to get a drug approved by the FDA, a company must obtain data from preclinical studies as well as clinical trials. Drug development is a very risky business: less than 30 percent of new drugs are profitable and very few drugs become blockbuster medications like Viagra or Prozac (Resnik 2001c; Goldhammer 2001).

The American Medical Association (AMA) recognizes the potential benefits of patents for medicine, although it also recognizes potential harms. The organization supports policies that are designed to encourage progress in science and medicine while protecting researchers, clinicians, and patents from some of the potential pitfalls of private control of DNA (Council on Ethical and Judicial Affairs 1997).

CASE STUDIES

In order to understand some of the potential problems that DNA patents can create for clinical medicine, I will describe five cases that have raised ethical concerns.

The Moore Case

Although this case does not involve a patent on DNA, it is worth reviewing because it draws attention to some potential ethical problems in the patenting of biological materials. In 1976, John Moore contracted a rare form of cancer known as hairy-cell leukemia and received treatment at the University of California, Los Angeles (UCLA) Medical Center. Dr. David Golde, Moore's physician, recommended a splenectomy. Moore consented to the treatment. After his spleen was removed, Golde asked Moore to make several visits to the Medical Center, during which time Moore was asked to provide samples of blood, skin, bone marrow, and sperm. Golde told Moore that these samples were needed to monitor his health but did not tell him that he was gathering the samples to develop a cell line from Moore's tissue. Golde was interested in Moore's tissue because it was overproducing lymphokines, which are proteins that play a variety of roles in regulating the cells of the immune system. The tissue had commercial value because it was a potential lymphokine factory. Golde and his research assistant, Shirley Quan, signed agreements with the University of California and several pharmaceutical companies to develop the cell line, which had an estimated value of $3 billion. Golde and Quan applied for and obtained patents on the cell line, which they assigned to the University of California (Gold 1996).

When Moore found out that he had been deceived, he sued Golde, Quan, the private companies, and the University for medical malpractice and for the tort of conversion, in other words the act of substantially interfering with his personal property. After several different appeals, the case reached the Californian Supreme Court, which ruled that Moore did not have a proprietary interest in his cells and could therefore not prove the tort of conversion (*Moore v Regents of the University of California* 1990, 156). However, the Court also concluded that Moore could be awarded damages for malpractice because Golde breached his fiduciary obligations to Moore by failing to disclose his commercial interests in Moore's cell line. The Court held that Golde had failed to provide Moore with adequate informed consent, because he had not told Moore about his commercial interests (*Moore,* 164). According to the majority opinion, only those people who isolated and cultured the cell line had property rights on the cell line. The majority also held that allowing patients to have property rights in their biological materials would undermine future R & D (*Moore,* 159). In order to promote medical progress, inventors and companies should have property rights over tissues, not patients. In two noteworthy dissenting opinions, Justice Broussard argued that Moore's tissue was his property because it was no different from other biological materials that are already exchanged on the market (Moore, 166) and Justice Mosk argued that Moore's tissue was his property because Moore should be able to have control over his own body and its parts (*Moore,* 173).

The Hagahai Case

Another case worth mentioning concerns the NIH's patent on viral genes taken from the Hagahai people, which we mentioned briefly in chapter 1. In

March 1995, the U.S. PTO awarded a patent to the NIH for a cell line infected with the human T-lymphotropic virus 1 (HTLV-1), which had been taken from the Hagahai people of Papua, New Guinea. The NIH pursued the patent in order to encourage companies to develop diagnostic tests for HTLV-1 variants and vaccines for fulminant leukemia, a disease sometimes caused by HTLV-1 infection. The NIH would grant an exclusive license to a company interested in developing a diagnostic test for the virus. In 1994, the NIH withdrew a similar patent on a cell line taken people in the Solomon Islands because it appeared that there was little commercial interest in products that might be developed from these cell lines, such as diagnostic tests. The NIH did not withdraw its Hagahai patent because Dr. Carol Jenkins, the physician who had worked closely with the Hagahai, asked the NIH to continue to pursue this patent. Dr. Jenkins hoped that a company would use the patent to develop a test for the virus, which she believed would benefit the Hagahai people. She had promised the Hagahai she would pursue this patent and give them a share of the royalties. However, this case set off an international controversy as advocates for indigenous peoples viewed it as another example of how western science and industry exploit people from the developing world. Over thirty political groups made formal declarations against the patenting of biological materials taken from indigenous peoples. These groups also opposed international attempts to understand the genetic diversity of the human population on the grounds that these efforts could lead to economic exploitation of indigenous people (Resnik 1999b).

The Canavan Case

On June 12, 1981, Jonathan Greenberg was born to Mr. and Mrs. Daniel Greenberg. When he was a year old, a neurologist diagnosed Jonathan with Canavan's disease which is a rare neurological disorder that occurs almost exclusively in Ashkenazi Jews, results from a deficiency in the enzyme aspartoacylase (ASPA). Jonathan died of the disease when he was 11-years old. Amy Greenberg, the Greenberg's second child, also developed the disease. The Greenbergs decided to lead an effort to identify the mutation that causes Canavan's disease, in order to insure that Jonathan's and Amy's suffering would not be in vain. The Greenbergs enlisted the aid of Dr. Reuben Matalon, a physician who was working at the University of Illinois Hospital in Chicago. The Greenbergs helped Matalon obtain blood, skin, and urine samples from children with Canavan's disease and their parents. They also raised about $100,000 in donations and grants for the project. Matalon was soon hired by Miami Children's Hospital (MCH) to establish a center for research on genetic diseases. MCH agreed to fund his work at a level of $1 million per year. In 1993, Matalon isolated the Canavan's gene and MCH applied for a patent on the gene, which the PTO awarded on October 21, 1997. In his contract, Matalon assigned all of his patent rights to MCH (Kolata 2000).

After MCH obtained the patent, it considered donating the patent to the public and forgoing any royalties, but eventually rejected this option because it feared that laboratories would not bother to publicize the test, since so few people have the disease. MCH decided to charge royalties of $12.50 per test to laboratories that perform the test, and planned to use the money from these fees to help offset the costs of R & D and publicity. MCH regarded $12.50 as a reasonable royalty fee for the test.

People in the Canavan's community did not agree with this opinion, however. They argued that laboratories should be able to perform the test without paying any royalty fees or signing any licensing agreements (Kolata 2000). On October 30, 2000, the Greenbergs and several other parties, including nonprofit groups, filed a lawsuit against MCH and Matalon in a Chicago federal court (*Greenberg v Miami Children's Hospital et al.* 2002). The lawsuit alleges six different cause of action, including breach of informed consent, breach of fiduciary duty, unjust enrichment, fraudulent concealment, conversion, and misappropriation of trade secrets. Canavan's families and organizations sought injunctive relief to prevent Miami Children's Hospital from restricting access to the genetic test and from impeding research on on Canavan's disease by enforcing its patent. The Canavan's Foundation had to stop its free genetic screening program because it lacked a license to perform the test (Kolata 2000).

THE BRCA1/BRCA2 CASE

During the 1990s, researchers from the United States and Europe discovered that mutations of specific genes known as BRCA1 and BRCA2 are associated with an increased risk of breast and ovarian cancer. Women with BRCA1 or BRCA2 mutations have an 87 percent chance of developing breast cancer by age seventy, as compared to a 10 percent chance for women from the general population. Women with the BRCA1 mutation have a 44 percent chance of developing ovarian cancer by age seventy, as compared to less than a 2 percent chance for women from the general population. BRCA1 mutations also increase the risk of breast cancer and prostate cancer in men. BRCA1 carriers have an increased risk of colon cancer as well. Even though BRCA mutations are associated with a higher risk of breast and ovarian cancer, less than 10 percent of breast cancer is associated with these mutations. There are over 235 different types of mutations that occur in BRCA1, and about 100 different mutations that occur in BRCA2 gene (Shattuck-Eidens et al. 1997; Kahn 1998). Myriad Genetics and the NIH both sponsored most of the research on BRCA1, and Myriad also claims that it was the first party to identify BRCA2. However, the European Breast Cancer Consortium and the Sanger Institute at Cambridge University have also conducted research on BRCA2 (Balter 2001).

During the 1990s, Myriad developed and patented genetic tests for BRCA1 and BRCA2 mutations. Myriad's assay identifies 16,500 base-pairs, or most of the known BRCA1 and BRCA2 mutations. This thoroughness has clinical significance, since a test that identifies 50 percent or less of these mutations might not provide useful information to patients. With its monopoly on the BRCA1 and BRCA2 test, Myriad sought to recoup its R & D expenses. The company charged $2400 for the test initially and granted licenses to only a dozen laboratories in the US in order to restrict the competition (Reynolds 2000). Myriad also stopped many unlicensed laboratories from performing the test. As a result of complaints about the high cost of its tests, Myriad allowed some laboratories to conduct the test for a lower fee. Some laboratories sacrifice thoroughness for economy by only testing for the most probable mutations, instead of all known mutations. Myriad has licensed the National Cancer Institute to test for BRCA mutations at a rate of $1,200 for BRCA1 and BRCA2, $600 for BRCA1 only, and $750 for BRCA2 only. In the future, Myriad is planning to expand its market to European countries although some companies have disputed its patent across the Atlantic (Foubister 2000). France, Belgium, and the Netherlands have challenged Myriad's European patents on the grounds that they are excessively broad in scope and grant the company a monopoly (Finkel et al. 2002).

The Gelsinger Case

The first gene therapy patient was a four-year-old girl with adenosine deaminase (ADA) deficiency who was injected with some of her own, genetically modified, bone marrow stem cells on September 14, 1990. The principal investigator in the experiment, French Anderson, had publicly defended gene therapy research for several years before the FDA gave him permission to begin clinical trials. Anderson's first two subjects showed improvement as measured by overall health and morbidity and laboratory tests of immune system function. While research groups have produced similar results, it is difficult to determine what percentage of positive treatment outcomes are due to receiving a drug known as PEG-ADA, which is part of the protocol. While gene therapy has not produced a cure for this disease, it offers ADA deficiency patients some hope for useful intervention (Walters and Palmer 1997).

During the 1990s, gene therapy research expanded. Several thousand patients have been enrolled gene therapy experiments, and the U.S. government alone has approved over 250 SGT protocols (Marshall 2000f). Researchers have conducted gene therapy clinical trials on combined immune deficiency (SCID), cystic fibrosis (CF), familial hypercholesterolemia, Canavan's disease, coronary artery disease, arterial restenosis, rheumatoid arthritis, Gaucher's disease, alpha$_1$ antitrypsin deficiency, Fanconi anemia, various forms of cancer, and HIV/AIDS. Most of the protocols approved in the United States are for cancer. Future clinical trials may target diseases such as Lesch-Nyhan syndrome, phenylketonuria (PKU), hemophilia A and B, Duchenne's muscular dystrophy, and Huntington's disease. Since 1990, public and private funding for gene therapy experiments has risen from several million dollars per year to several hundred million dollars per year. The Patent and Trademark Office has awarded sixty-four gene therapy patents (PTO 2002).

Gene therapy research suffered a key setback when Jesse Gelsinger died in a SGT trial at the University of Pennsylvania on September 17, 1999. Gelsinger was receiving a treatment to correct a disease that occurs when patients produce insufficient quantities of a liver enzyme known as ornithine transcarbamylase (OTC), which helps the body remove ammonia from the blood. The disease results from a mutation of a gene that codes for OTC, which is located on the X chromosome. Patients with the disease develop ammonia toxicity and often die at birth. Patients with milder forms of the disease, such as Gelsinger, can continue to live if they follow a strict diet and take medications to reduce blood ammonia levels. In 1994, James Wilson, Steve Raper, and Mark Batshaw had designed a protocol for transferring an OTC gene to patients in order to replace the defective gene. The protocol used a safe adenovirus as a vector to deliver the gene. The research team's initial plan was to test their treatment on infant patients who would be on the verge of death from ammonia toxicity (Marshall 2000f).

In 1995, Arthur Caplan, a bioethicist at the University of Pennsylvania consulted with Wilson, Raper, and Batshaw and convinced the team to first give the treatment to a healthy subject, in order to test toxicity levels. Thus, Gelsinger was recruited as a patient for the trail because he was relatively healthy at the time. Caplan also argued that since the research would have a very low chance of providing a benefit to the patient, it should be conducted on an adult, not a child, since a child would be unable to consent to the research. Batshaw also discussed this revision of the plan with the National Urea Cylce Disorders Foundation and the Recombinant DNA Advisory Committee. Both organizations agreed with Caplan's recommendation. In 1997, a safety review by the FDA recommended that injecting the adenovirus directly into the liver would pose less of a risk to patients, because it would be less likely to cause a systemic immune response. Gelsinger therefore was given injections of the adenovirus in his liver. He died when he developed a severe immune response to the adenovirus vector that the researchers did not anticipate. An autopsy of Gelsinger revealed that the virus did not remain in his liver but had spread throughout his body. Researchers are rethinking future plans to use adenoviruses as vectors in gene therapy and are looking to safer vectors (Marshall 2000f).

Since medical research provides subjects with potential benefits as well as risks, it is not unusual for patients to die during clinical trials. Indeed, when a patient's prognosis is grim, as in advanced liver cancer, the mortality rate may be 90 percent or higher. However, Gelsinger's death caused a stir because his prognosis was not grim and gene therapy is in the media's spotlight. The case made national headlines and helped to instigate additional scrutiny into the ethical and financial aspects of clinical trials as well as proposed reforms of regulations pertaining to conflicts of interest, data and safety monitoring, and institutional review boards. The U.S. government has not created any new regulatory agencies to monitor gene therapy, although it has placed more scrutiny on gene therapy clinical trials.

One of the important ethical and financial issues that emerged during the investigation and discussion of Gelsinger's death concerned the conflicts of interest (COIs) of the investigators and the institution. Wilson and the University of Pennsylvania held equity in Genovo, a private firm founded by Wilson. Genovo had sponsored the trial in which Gelsinger died and was providing $4 million per year to the university's Human Gene Therapy Institute. Wilson also holds patents on several of the gene therapy techniques used in the experiment and could realize profits his gene therapy inventions (Marshall 2000f). Thus, both Wilson and the university had financial interests in the clinical trial. Even if these financial interests did not affect the quality of the research, they certainly created the perception of bias.

The case also involved a lawsuit. The Gelsingers sued James Wilson, the University of Pennsylvania, Genovo, Inc., Steve Raper, Mark Batshaw, William Kelley, the Children's Hospital of Philadelphia, Children's National Medical Center, and Arthur Caplan (Gelsinger Complaint 2002). The suit alleged wrongful death, assault and battery, products liability, lack of informed consent, intentional and negligent infliction of emotional distress, and fraud. The parties settled the case out court for an undisclosed amount of money.

EXPLOITATION IN GENETIC RESEARCH

One of the main problems raised by these cases concern the potential for the exploitation of patients/subjects. In the most egregious example, the Moore case, the researchers and the company had a potential profit of billions of dollars but they offered Moore no money for his tissue. They also did not inform him about their plans to use his tissue. In the Hagahai case, it does not appear that subjects were exploited, but many people believed that they (and other indigenous populations) would be exploited by future patenting activities in the developing world. Exploitation was also an issue in the Canavan case, since one might argue that MHC exploited the Greenbergs and the Canavan community. Finally, the Gelsinger case also involved exploitation, if one holds that the researchers exploited Gelsinger in order to secure his participation in a clinical trial.

Before discussing the topic of exploitation in more depth, it will be useful to develop a working definition of the concept. To exploit a person is to take unfair advantage of that person (Wertheimer 1996). Exploitation can occur when one person exploits one person, when one person exploits a group of people, when a group exploits one person, or when one group exploits another group. When exploitation occurs in medical research or practice, it is usually when a group of researchers or professionals exploits a single person or a group of people. There are several characteristics of exploitation. Practices that we call exploitative have some or all of these characteristics. Usually, the most egregious cases of exploitation in research, such as the Tuskegee syphilis study or the Nazi experiments, have all of these characteristics (Resnik 2003b). These characteristics do not provide necessary and sufficient conditions for defining exploitation, but they are features that we commonly recognize as part of the concept of exploitation (Wertheimer 1996; Resnik 2003b):

1. Harm. Frequently, exploitative acts cause harm to the exploitee.
2. Violation of dignity. Frequently, exploitative acts violate the dignity of the exploitee by treating the person as a mere means to another end.
3. Inadequate informed consent. Frequently, exploitative acts involve deception, fraud, duress, coercion, undue influence, or taking advantage of a person's vulnerabilities.
4. Injustice. Frequently, exploitative acts involve inequitable or imbalanced transactions between people.

For example, let's consider the Moore case again. In this case, Moore's informed consent was inadequate because the researchers did not tell him some important information he needed to know to make a decision. They did not disclose information to him about the value of his tissue or their financial interests. They violated Moore's dignity, because they treated him as a mere means to their own ends (e.g., profit and career advancement). The transaction between Moore and the researchers was also inequitable because Moore received no compensation for his tissue while researchers stood to earn millions of dollars. Finally, although we do not have any evidence that the researchers caused Moore actual physical harm, it is likely they inflicted psychological harm on him.

It will also be useful to distinguish between actual exploitation and the perception of exploitation, since people may feel that they are being exploited even though they are not really being exploited. For example, when gasoline prices rises in the summer months, many people may feel that the oil companies are exploiting their customers through price gouging, even though the prices for gasoline may accurately reflect the current balance of supply and demand. Although creating the perception of exploitation is not unethical in itself, it can still have negative effects on those people who feel that they are being exploited and can undermine trust. In medicine, patients may lose their trust in doctors if they believe that they are being exploited for financial or other reasons. Trust is an essential part of the doctor-patient relationship in that it plays a key role in honest an open communication that is necessary for effective diagnosis, treatment, and prevention. Although it is not possible to prevent all patients from feeling that they are the victims of exploitation, it is important to avoid creating the perception of exploitation when it is feasible to do so.

Clearly, the threat of exploitation posed by DNA patents is credible threat. The Moore case is not an imaginative nightmare scenario. Moore was exploited in a most egregious way (Andrews and Nelkin 2001). Although the Moore case did not involve a DNA patent, the same financial interests that led researchers to exploit Moore for his cells could lead others to exploit other patients for their DNA. Moreover, the threat posed by the perception of exploitation is also a credible threat to medicine. Several decades ago, patients would trust their doctors unconditionally and would have little knowledge or concern about their financial interests. Today's patients, on the other hand, are less trusting and are more concerned about the financial and commercial interests of physicians, medical researchers, universities, and private corporations. It is important to try to dispel patients' fears about exploitation and affirm trust in medical research (Korn 2000).

What is a reasonable response to threat of exploitation, both actual and perceived? Although the threat is plausible and disturbing, it is not so dangerous or unique as to justify a ban on DNA patents. Two arguments support this conclusion. First, it is not likely that many patients will have DNA samples that have a tremendous amount of commercial value. Moore had a unique mutation. Most of the commercially valuable information in human genetics research will come from studying populations of individuals, such as the Hagahai, Ashkenazi Jews, the Icelandic people, or people with heart disease (Annas 2000). Thus, it is not likely that the kind of exploitation that happened in the Moore case will be repeated very often. That being said, it is still important to take steps to protect populations from exploitation.

Second, patenting plays a major role in the development of virtually all drugs and medical devices. Companies and researchers hold patents on medications, preparations, formulas, diagnostic tests, diagnostic equipment, laboratory equipment, medical information systems, and artificial body parts. DNA patents are only another drop in the bucket when it comes to medical patenting. Banning DNA patents would therefore have little effect on the influence of patents in medicine and the potential exploitation of patients. One would have to ban all biotechnology patents, including patents on cell lines and proteins, to effectively counter this threat.

Although banning DNA patents is not the best course of action, there are some prudent and precautionary measures that physicians and researchers can take to guard against exploitation and the perception of exploitation. First, researchers and physicians should follow accepted standards of informed consent in genetic research. Informed consent is required when genetic samples can be linked to individuals, but it is not required for anonymous samples (Clayton et al. 1995). Researchers and physicians should disclose all material information to patients, including the nature of the research, benefits and risks of research, and other options (National Bioethics Advisory Commission 1998; McQueen 1998). Researchers should also disclose their financial interests in research as well the potential financial impact of the research (Korn 2000). The researchers in the Moore case did not live up to this standard, but other medical researchers should. Any researcher who takes tissue samples from patients in order analyze their DNA should explain to those patients their own financial interests in the research project as well as the potential financial impact of the research. The same principle of disclosure would apply whether the patients are Ashkenazi Jews living in New York City or Hagahai people living in Papua, New Guinea.

Second, to protect communities from harm and injustice, researchers should seek consent from the community as well as from individuals, in some circumstances. Community consent is appropriate when a community of people is likely to suffer discrimination, bias, or exploitation as a result their participation in research. Community consent should involve consultations with community leaders and representatives, as well as informal discussions with other members of the population. Community consent should supplement, but not replace, individual consent (Sharp and Foster 2001).

Third, researchers, physicians, and companies should offer patients or populations fair compensation for their contributions to genetic research. We should avoid the situation encountered in *Moore,* where the research subject received no money for his tissue samples but the researchers and the company stood to make millions of dollars. If researchers discover a DNA sequence in a human being that has tremendous commercial value, then they should give that person fair compensation for that sequence. What constitutes a fair price for a tissue or DNA sample is a question I will not try to answer, but the price should reflect the market value of the DNA patent and should be reached through good faith negotiations. If an individual is not interested in money, he or she should be offered some other form of compensation.

However, the Moore case is an exception rather than rule, because most individuals who participate in research will not make contributions that have a high or even measurable market value. Most of the DNA sequences collected in research will only have commercial value in relation to a large set of samples collected from a population. Thus, a DNA database may have great commercial value, even though individual samples do not. As an alternative to compensating individual patients for their DNA, researchers should consider providing some compensation to the population. This compensation could range from a discount on DNA tests or products, to contributions to a research or education fund, to the provision of education or health care. For example, MCH has attempted to provide compensation to the Canavan's community by licensing the test at a nominal fee and providing more money for research. In the Hagahai patent, researchers offered to give the Hagahai a share of profits from the patent. The issue of providing populations with the benefits of research participation is not unique to genetic research and has arisen in other contexts, such as the conduct of clinical trials in the developing world. The argument that researchers should compensate populations is grounded in notions of justice and fairness in research, which play a prominent role in The Belmont Report (National Commission for the Protection of Human Subjects 1979) and The Declaration of Helsinki (World Medical Association 2001).

Admittedly, exploitation and the perception of exploitation in research may still occur no matter how many precautions society takes. However, the three precautions discussed above will help avoid these problems and promote fair treatment of research participants.

CONFLICTS OF INTEREST IN MEDICAL RESEARCH AND PRACTICE

As we saw in chapter 7, conflicts of interest (COIs) can undermine the objectivity and trustworthiness of biomedical research. They can also affect the practice of medicine (Rodwin 1993). A conflict of interest in clinical medicine is a situation in which a physician's personal or financial interests undermine professional judgment and decision-making. For example, if a physician has a 20 percent ownership interest in a medical laboratory, he or she will have a financial incentive to send patients to the laboratory for medical tests. As a result of this financial interest, the physician might recommend unnecessary medical tests. If a physician receives gifts from a pharmaceutical company, such as expensive pens, free lunches, and trips to resort hotels, he or she may be inclined to prescribe that company's medications even when the prescription would not be in the patient's best interests. Many of the COIs that abound in medicine provide physicians with incentives to overtreat their patients because physicians have tended to make more money when they do more things for patients. The advent of managed care in the 1990s has given physicians financial incentives to under-treat patients, because managed care organizations reward physicians who practice cost-effective medicine and punish those who do not.

Physicians and surgeons who treat patients with medications and therapies may also have financial interests that could compromise their judgment and decision-making. For example, a physician with a patent on a device might use deception and coercion to recruit patients for clinical trials in order to test his new medical technology. A physician with a patent on a medication might be inclined to prescribe that medication to his patients in order to advance his own financial interests. When a physician has a DNA patent, similar financial interests could motivate his or her conduct: he might recommend an unnecessary DNA test or DNA-related therapy, such as gene therapy or aggressively recruit patients into clinical trials. COIs were a major concern in the Gelsinger case, since many people believed that Wilson's financial interests related to his gene therapy patents could have affected his conduct. We will probably never know whether Wilson's COIs affected his judgment and decision-making. Would he (and the other researchers) have been more cautious if they had not had a financial stake in the experiment? Would they have not recruited Gelsinger as a subject? Regardless of whether Wilson, other researchers, or the University of Pennsylvania had a genuine COI, their financial interests created the appearance of a COI. It is important to deal with problems relating to apparent COIs, since apparent COIs can undermine trust even when they do not affect objectivity (Resnik 1998b).

Clearly, all patents held by physicians pose a serious threat to the practice of medicine. This threat applies to DNA patents as well as patents on new drugs, medical devices, procedures, and diagnostic tools. Thus, DNA patents do not pose a uniquely dangerous threat, because the threat already exists. If one follows the Precautionary Principle, one should ask, What is a reasonable response to this threat? As noted in chapter 7, there are several different strategies for dealing with COIs, including disclosure, management, and prohibition and avoidance. Some writers (e.g., Merz and Cho 1998) have argued physicians should not hold DNA patents because the patents create unacceptable COIs or apparent COIs. These writers are concerned that patent ownership will bias or cloud the physician's medical judgment and threaten patient safety and doctor-patient trust. To assess this critique, we should consider consequences of banning medical patents: if individual physicians are not allowed to hold patents, then who will? Who would apply for medical patents and develop new drugs and medications? The only reasonable alternative would be that some outside party, such as a university or a pharmaceutical or biotech company, would hold patents and would develop new medications. To make this situation work, practicing physicians would almost need to be barred from conducting any medical research, since medical research frequently leads to patentable products. Thus, an outside party would design and develop medical inventions; physicians would simply test them on patients for this third party. Now we can easily see some of the difficulties with banning medical patents. First, this suggestion is very impractical and unrealistic because to make it work one would need to prevent physicians from conducting medical research. At most, they could test inventions for someone else, but they could not design and develop inventions. Second, since this suggestion would require a third party to be a part of the physician-patient relationship whenever medical research is conducted, it is doubtful that this situation would improve patient safety or doctor-patient trust, because the third party might develop incentives that would tend to promote its own financial interests but could bias the physician's medical judgment. For example, a pharmaceutical company might pay physicians for collecting data. Third, allowing physicians to hold medical patents can improve patient care because patenting provides physicians with incentives and rewards for medical innovation and scientific discovery (Council on Ethical and Judicial Affairs 1998). The reward mechanisms that justify patents for scientists also justify patents for physicians as well as other health care professionals.

Although patents create COIs for physicians, it would not be wise to prohibit physicians from holding patents. A reasonable response to the threat posed by medical patents would be to pursue other strategies for dealing with COIs, such as disclosure and management. Disclosure and management can help to guard against bias by allowing outside parties to observe and review the COI. These strategies can also help to promote trust by encouraging open communication about financial interests instead of secrecy and evasion. Physicians should disclose their financial interests to patients, including any interests they may have related to DNA patents. Physicians, medical schools, professional associations, and funding agencies should develop and implement plans and policies for managing COIs in medicine and medical research, and should form committees to review and monitor COIs (Korn 2000). Thus, although COIs related to DNA patents pose a plausible threat to the physician-patient relationship, the most reasonable response to this threat is to take precautionary measures, such as disclosure and management of COIs, instead of banning DNA patents altogether.

THE COMMERCIALIZATION OF MEDICINE

A problem closely related to exploitation and COIs is the contribution of DNA patents to the continued commercialization of medicine. Many commentators have lamented the role of financial and commercial forces in transforming medicine from a profession into a business (Hanson 1999; Churchill 1999). According to the professional approach to the physician-patient relationship, doctors have ethical duties to promote the interests of their patients and protect them from harm. As professionals, doctors must often place their patient's interests ahead of their own. These ethical duties are based on the physician's role as a fiduciary for the patient. As fiduciaries, physicians have expertise and knowledge that patients usually lack. Patients often seek the care of a physician when they are sick and therefore physically and psychologically vulnerable. The patient therefore depends on the physician to use that knowledge and expertise wisely and ethically. Trust plays a very important role in this professional model of the physician-patient relationship because patients must trust their doctors to provide them with care and expertise (Wynia et al. 1999).

Financial and economic forces are changing medicine from a profession into a business. First, doctors have adopted a business model for the management of patients and medical services. Many physicians now have contracts with managed care organizations that specify rules for reimbursement for services and provide incentives to control costs (Morreim 1995). These rules and incentives encourage physicians to make medical decisions based on cost/benefit analyses rather than the patient's best interests. Public and private insurers also place economic restrictions on physicians. Due to increased costs of care and lower reimbursement rates, physicians are also less likely to provide services for free or at a lower cost. In managing their own private practices, physicians often make decisions based on the bottom line, rather than on what's best for patients. Physicians also have financial interests that can threaten their duties to patients, such as referral arrangements, consulting fees, ownership of medical companies, stock, and patents (Rodwin 1993).

Second, as a result of the decline in paternalism in medicine and a rise in the emphasis on patient autonomy, many patients have adopted a consumer mentality and often regard doctors as health care providers or suppliers (Hanson 1999). Patients frequently expect to make choices about a wide variety of treatment regimens and alternatives and they expect doctors to meet their demands. Consumerism in medicine abounds in cosmetic surgery, assisted reproduction, and the demand for drugs designed to affect personality and emotions, such as Prozac. Patients want to have a great deal of choice in health care and they demand quality and accountability. Pharmaceutical companies encourage this consumer mentality by marketing drugs directly to consumers through advertisements in the media. When something goes wrong, they are more apt to sue the physician.

Finally, the entire culture of medicine has become more businesslike. Large corporations, private insurers, hospitals, managed care companies, and government organizations, such as Medicaid and Medicare, now dominate medical practice, healthcare decision making, and health care policy. Frequently, decisions made in a corporate boardroom or government committee have more of an impact on health care than decisions made at the bedside. Healthcare, especially hospital care, has become less personalized and more institutionalized. Patients often do not know the names of their healthcare providers and there is a constant turnover in the medical and nursing staff from one shift to the next.

There is no need to dwell on the various ways in which medicine is becoming more like a business. I think few people would dispute this claim.

If we recall the discussion of commodification in chapter 6, this critique of modern medicine can be construed as calling our attention to the fact that medicine is an incomplete commodity that is in danger of sliding down the slippery slope toward complete commodification (Hanson 1999). Patenting in health care can contribute to the further commodification of health care. Therefore, one might argue, the patenting of DNA threatens medicine because it adds to the commodification of medicine.

Clearly, the threat of further commodification in medicine is plausible. The changes in medicine in the last century are dramatic and well documented. Although this is a real problem and concern, we have little evidence for thinking that DNA patents have made or will make a significant contribution to the commodification of health care. All of the changes described in this section would have occurred even if no one had ever patented DNA. DNA patenting may make a minor contribution to the trend toward commercialization in healthcare, but its impact is small in comparison to the impact of managed care, medical insurance, physician's fees, the direct marketing of medicine to consumers, and other changes in health care during the last 100 years.

Since DNA patents have had a minimal impact on the transformation of medicine, banning DNA patents would certainly do little to address the threat of further commodification. The best course of action is to take precautionary and proactive measures to promote professionalism in health care and prevent medicine from becoming merely a business. Some of these measures include rules pertaining to contractual and financial relationships in medicine, disclosure rules, reimbursement rules, ethical guidelines, and so on. The topic of how to promote professionalism in medicine is beyond the scope of this book, so I will not discuss it further here (see Wynia et al. 1999).

ACCESSIBILITY TO GENETIC MEDICINE

The final threat to medical practice posed by DNA patents concerns accessibility to genetic medicine. According to this critique, the patenting of DNA will drive up the price of genetic tests, gene therapy, and genetic treatments and will therefore prevent many people from obtaining access to genetic medicine. Critics of DNA patents cite Myriad's high licensing fees for its BRCA1/BRCA2 test as proof of this problem (Merz 2000). One might add that the high costs of drugs derived from genetic research, such as Epogen or synthetic clotting factors, also pose access problems for patients. If companies follow Myriad's example, only patients with a lot of money or very good health insurance will be able to afford genetic tests and the gap between the haves and have-nots in health care will grow. Thus, medical patents drive up the cost of medical procedures and medications, which decreases access (Garris 1996).

In responding to this criticism, I will assume that principles of justice and fairness require us to promote accessibility to health care, including access to genetic medicine (Daniels 1984; Mehlman and Botkin 1998).[1] The question we need to consider is whether DNA patents are likely to decrease or increase access to genetic medicine, in the long run. I will assume, for the sake of argument, that cost will be the prohibitive factor in access to genetic medicine in the short-term. So, we need to ask whether patents enhance or undermine access to health care, in the long run.

Incidentally, there are other cases where we accept short-term injustices in order to serve justice in the long run. For example, the U.S. legal system has a variety of rules designed to protect the rights of someone accused of a crime. Very often, enforcement of these rights leads to injustices in the short-term. A murderer might be allowed to roam the streets because the police violate rules for obtaining evidence or they conduct an unlawful search and seizure. In the short-run, it might seem to be very unjust that the murderer is allowed to go free, but in the long run justice is well-served by having rules that protect the rights of the accused, since these rights make it less likely that innocent people will be convicted and they protect basic constitutional liberties (Israel and LaFave 1993).

Clearly, the patent system can create short-term problems with access to technology because it allows inventors to have exclusive control over their inventions for a set period of time. During the life of the patent, inventors have a monopoly on their invention and can charge the high prices that are possible under this form of control. Prices are inflated because patent holders control the supply of the product or process. In a free market, laws of supply and demand determine prices; the price of a good or service will reflect what consumers are willing to pay, given the supply. If a company can control the supply of a good or service via its monopoly power, and the demand remains constant, then the company can control the price. If a company loses its ability to control the supply and competitors enter the market, prices will decrease as a result of an increased supply (Samuelson 1980). Myriad Genetics can charge a high price for its test because it has a monopoly on the test and consumers are willing to pay its price.

Once a patent expires, prices will go down as other competitors enter the market. For example, pharmaceutical companies can charge high prices for their drugs during the life of their patents. Once a patent on a drug expires, another company can introduce a generic version at a lower cost. Prices will also drop as a result of technological innovations that are driven by economic forces. Very often, a company will be able to enter the market for a particular good or service by developing an improvement upon an existing good or service or an entirely new good or service. For example, if a pharmaceutical company, Company A, patents a blood pressure medication, another company, Company B, may be able to enter the market for blood pressure drugs during the life of the patent owned by Company A, because Company B has developed an improvement on the invention owned by Company A or as has developed a new invention. The process of research, product development, patenting, competition, innovation, research product, development, patenting, and so on, reiterates many times during the development of a new industry. As a result, prices go down, access goes up, and the technology improves. Many technological innovations, such as automobiles, personal computers, and televisions were very expensive and unaffordable when they entered the market but came down in price as a result of further innovation, competition, and economies of scale (Volti 1995). Since patents encourage technological innovation and competition, they generally help to bring down the price of a product in the long run, even if they increase its price in the short-term. Without patents, companies would either not invest in R & D or they would resort to trade secrecy to protect their proprietary interests. We have no reason to believe that either of these options would promote the progress or science or fair competition. So, patents tend to promote the accessibility of technology in the long run (Miller and Davis 2000).

On the other hand, this argument assumes that there are no conditions or forces that disrupt market competition. Monopolistic control of an entire market or industry by a single company, collusion among several companies, and inadequate consumer knowledge and choice can interfere with free market competition. In patenting, problems can arise as a result of patents that are overly broad. Although patents are designed to give the inventor a limited monopoly on his or her invention, they should not be so broad that they prevent competitors from entering the market with new and improved inventions. For example, if a company develops a diagnostic test for a genetic condition, other companies should still be able to enter the market for genetic testing by developing improvements on the original invention or entirely new inventions. Patenting agencies and the courts should give careful consideration to the appropriate scope of patents by balancing the need to promote fair competition against the need to provide adequate incentives to inventors. Problems with fair competition can also arise through various abuses of the patent system discussed in chapter 3, such as double-patenting, using a patent to block technological development in an area, and attempting to wrongfully extend the scope of a patent. In some cases, it may be appropriate to use antitrust law to deal with unfair patenting practices (Lao 1999).

Thus, while the patent system tends to promote access to technology in the long run, it is not perfect. In order to promote access to genetic medicine, patenting agencies and the courts need to develop and implement policies that limit monopolistic control of DNA inventions. Implementing these policies requires patenting agencies and the courts to balance public vs. private interests in the following areas:

1. Scope. DNA patents should not be so broad in scope that they prevent fair competition from other inventions, including improvements.
2. Utility. DNA patents should be based on the demonstration of a definite utility.
3. Life of the patent. The life of a patent should not extend beyond twenty years. Agencies and the courts should prevent double-patenting and other tactics designed to keep competitors off the market.
4. Antitrust law. If other methods of promoting fair competition fail, federal agencies may use federal antitrust laws to promote fair competition (*United States v White Motor Co.* 1961). Although patents give patent holders a monopoly on a particular product, they should not give the patent holder control over an entire market. If a patent holder attempts to wrongfully extend the scope of his patent, the U.S. legal system has mechanisms in place to deal with this type of abuse. (Antitrust law is a broad area of law beyond the scope of this book. See Gellhorn and Kovacic 1994)

Applying these points to Myriad's patent on BRCA1 and BRCA2 testing for cancer predispositions, I would stress that the scope of its patent should not be so broad that other inventors cannot enter the market to challenge Myriad's product. For example, patent agencies should award patents for tests that use different DNA sequences (or different combinations of DNA sequences) to test for cancer predispositions. Patent agencies should also award patents that use proteins, RNA, or other molecules to test for cancer predispositions. Even a test for some of the BRCA mutations could be patented because it could represent a more cost-effective (and therefore improved) way of testing for hereditary breast cancer. The same points would also apply to patents on gene therapy techniques, gene therapy vectors, and drugs produced through genetic manipulation. Myriad should not be able to extend the life of its patent through double-patenting or other abuses of the patent system. If Myriad were able to gain monopolistic control over the entire genetic testing market then federal agencies and regulators would be justified in taking steps to break up or weaken its domination of the market.

Thus, while the threat to accessibility to genetic medicine posed by DNA patents is plausible, banning DNA patents would not be a reasonable response to this threat. Indeed, banning DNA patents would probably not increase access to genetic medicine, since it would seriously impair incentives for technological innovation and private investment in genetic R & D. The most reasonable course of action is for agencies to administer the existing patent laws to encourage competition and to prevent abuses of the system.

Before concluding this section, I would like to address an important exception to the policy defended above (i.e., the health emergency exception). If a country faces a health emergency, such as the HIV/AIDS crisis in sub-Saharan Africa, then it may take steps to override or restrict patents, provided that it also attempts to provide fair compensation to patent holders. Under these extreme conditions, societies may set aside long-term goals in order to promote short-term ones. Under the TRIPS agreement, discussed in chapter 3, a country may use compulsory licensing to deal with a public health emergency (Resnik and De Ville 2002). For example, if a country needs a particular patented medicine, it could require the patent holder to license other companies (or the government) to make the medicine in order to increase the supply of the medicine and reduce costs. As an alternative to compulsory licensing, companies can subsidize the prices of medications, and companies can cut their prices or give away medication. In applying this policy, it is important for countries to develop a clear and precise definition of a health emergency, since a vague definition could open the door to abuses and undermine the patent system (Resnik and De Ville 2002). Countries should also enter into good faith negotiations with patent holders and cooperate with companies (Resnik 2001c). (I know of no situations in genetic medicine where a health emergency has occurred, but it is a theoretical possibility.)

ARE PATENTS ON GENETICS TESTS PATENTS ON PRODUCTS OF NATURE?

Before concluding this chapter, I would like to discuss an objection to patents on genetic diagnostic tests raised by Merz ands Cho (1998). According to this objection, the diagnostic tests developed by Amgen should not be patentable because they attempt to obtain exclusive control over natural regularities, which are products of nature. The regularities that these patents attempt to control are the associations between the presence of specific genes or mutations and predispositions to develop cancer. The tests take advantage of these statistical relationships to provide patients with information about their predispositions. You cannot patent a statistical association any more than you can patent laws of nature or other natural phenomena. Thus, you cannot patent genetic diagnostic tests.

To see the flaws in this reasoning, let's consider carefully how one might patent a diagnostic test in genetics. First, one could obtain a product patent on the DNA sequences themselves. The patent would be on an isolated and purified DNA sequence; it would not be a patent on a natural DNA sequence. The utility of the sequence would be to test for cancer (or other) predispositions. Thus, this type of patent would not make any claims on statistical associations. Second, one could obtain a process patent on methods for conducting genetic diagnosis. Such a patent would assert a claim over a particular method for diagnostic testing. The claim would describe the processes of isolating and purifying the genes that predict phenotypic outcomes. If this type of patent is drafted and reviewed correctly, then it should not give the inventors proprietary control over the statistical association itself: they would only be able to patent an application of that association in a particular type of invention. In theory, someone else could use that same association to develop a test that is different from or improves upon the prior, patented test. For example, another inventor could patent a test that uses a different method for detecting DNA sequences; another inventor could patent a method the uses a gene product, such as RNA or proteins, to test for predispositions. The general point here is that a patent on a diagnostic test should not give the inventor any control over a statistical association; other researchers should be able to use that association to develop new inventions. If someone actually obtained a patent that gave them control over a statistical association, then an outside party should challenge that patent as unlawful. This problem provides us with one more example of the importance of setting the proper scope for a patent. In this case, the scope of a patent on a diagnostic test should not be so broad that it effectively gives the patent holder a right to control a statistical association. The scope should be narrow enough that other inventors can also use that association without fear of infringing patents.

Summary

This chapter has considered several threats to the practice of medicine posed by DNA patents. It has argued that while these threats are plausible, they are not so serious as to justify a ban on DNA patents. The most reasonable course of action is to safeguard medical practice and research by developing and adopting policies designed to prevent exploitation, the erosion of trust, and promote professionalism and fair competition.

9

DNA Patents and Agriculture

INTRODUCTION

This chapter continues in the theme of the benefit-harm analyses conducted in chapters 7 and 8 but will focus on the potential benefits and harms to agriculture resulting from DNA patenting. After reviewing the benefits of DNA patents for agriculture, the chapter will examine the potential harms. As in chapters 7 and 8, the main concern will be to determine whether these harms are plausible threats, and if they are, what a reasonable response would be to them.

THE BENEFITS OF DNA PATENTS FOR AGRICULTURE AND AGROMEDICINE

Before discussing the potential harms to agriculture that DNA patents may bring about, it will be useful to briefly review the potential benefits of patenting, since benefit-harm decisions must balance both benefits and harms. The potential benefits of DNA patents are all related to the effect that patents have on basic and applied research, since research is essential to improvements in agriculture. Research that has an impact on agriculture includes basic research disciplines such as genetics, genomics, proteomics, biochemistry, and microbiology, as well as applied research disciplines such as genetic engineering, agronomy, plant pathology, entomology, animal husbandry, veterinary medicine, and nutrition. Although the genetic revolution has not yet had a major impact on medicine, it has had a profound impact on agricultures since the

1980s, when scientists developed genetically modified (GM) crops and animals. The first GM organisms, such as Chakrabarty's bacterium, were plants. In the 1980s, researchers developed GM crops and animals, such as tomatoes, corn, and rice, as well as mice, cows, and sheep (Rollin 1995).

There are several benefits of GM crops. First, genetic modification can enhance crop yields. Genetic engineers can design plants to produce more fruit and edible matter. Crops can also be genetically fortified with important vitamins for better nutritional value, such as Monsanto's beta carotene rice. More than half of the improvements in the yields of cereal crops since 1930 are due to the application of conventional breeding techniques (i.e., the selective breeding of hybrids). Conventional breeding techniques have also accounted for 85 percent of the improvements in the yield of soybeans (Shoemaker et al. 2001). However, selective breeding techniques are far less efficient than genetic engineering at increasing agricultural productivity: it may take dozens of years to develop a new hybrid via conventional means but only a year or so using gene transfer techniques.

World demand for cereal crops, such as corn, wheat, and rice, will increase 40 percent by 2020 (Shoemaker et al. 2001). It is estimated that the world population will grow to 10 billion by the year 2030 (Biotechnology Industry Organization 2001a). Almost all of the population growth will occur in developing nations, since industrialized nations are estimated to grow at a rate of less than 1 percent per year. The total acreage of agricultural land is expected to grow at less than 0.3 percent per year (Shoemaker et al. 2001). Unless the world takes effective measures to control population growth and increase the food supply, famine, starvation, and malnutrition will continue to get worse in developing nations. Although increasing crop yields through genetic manipulation is not the only, or even the most important strategy, for dealing with the problem of world hunger, it can play an important role in increasing the food supply (Wambugu 1999).

Second, genetic modification can reduce the need to use pesticides, fungicides, and herbicides to control insect and plant pests. Modern farmers use a variety of pesticides, fungicides, and herbicides on crops in order to reduce crop losses due to insects, pathogens, and weeds. These chemicals, though effective, can contaminate the soil, water, and ecosystem and thereby threaten the environment and public health. Researchers have developed GM crops that minimize the use of pesticides, fungicides, and herbicides. Some GM crops that have been developed are highly resistant to highly effective herbicides, such as Round-up, which allows the farmer to kill weeds without killing his crop. Other GM crops have been developed to secrete or contain pesticides, allowing the farmer to use fewer pesticides in the field. For example, researchers have developed plants that contain a protein produced by the bacteria *Bacillus thuringiensis* (BT), which is toxic to insects but harmless to other animals and people (Biotechnology Industry Organization 2001a).

Third, it may also be possible to develop GM crops that grow under harsh conditions where other crops cannot grow, such as arid climates, hot climates, or cold climates. By developing crops that can grow under different environmental conditions, genetic engineers can increase the percentage of arable land (Shoemaker et al. 2001). It is especially important to develop drought-resistant crops to serve the needs of nations in the developing world that have been stricken by years of drought and famine, such as Afghanistan, Ethiopia, Somalia, and Sudan (Wambugu 1999).

Genetic modification of plants and animals may also have important impacts on medicine. Bioengineers are also developing plants to produce human hormones and other medicines. For example, researchers at the biotech company Agracetus have developed a variety of GM corn that produces human antibodies that can be used to treat cancer. Another biotech firm, Croptech, has inserted genes into the tobacco plant that allow it to produce an important human enzyme known as glucocerebrosidase, and Plant Biotechnology is using GM tobacco plants to manufacture medicines that prevent tooth decay (Gibbs 1997).

Current research on genetic engineering and cloning in mammals may enable researchers to harvest organs from pigs to transplant into human beings. Normal pig organs produce antigens on their cell surfaces that trigger a major immune response in human recipients. The antigens are proteins that are coded for by a set of genes. Researchers are developing ways to deactivate (or knock out) these genes in pigs so their organs will be immunologically compatible with the human body (Chea 2002). Ian Wilmut, the scientist who produced Dolly, the first cloned sheep, is developing methods for transferring human genes into sheep to allow the sheep to produce human hormones, enzymes, and other important chemicals in their milk (Wilmut 1998).

Finally, since the 1980s, GM mice have played an important role in biomedical research by serving as models for many different human diseases and conditions, such as cancer, diabetes, obesity, anorexia, hypertension, Parkinson's disease, and heart disease. A common technique for developing a mouse model for a human disease is to deactivate a gene whose mutation is linked to a specific disease. For example, the P53 gene encodes a protein that functions to suppress tumors, and mutations of this gene can cause cancer. Some GM mice used in cancer research lack functional tumor suppressor genes, such a P53 (Malakoff 2000). Scientists have also transferred genes into rhesus monkeys as a first step toward developing GM monkeys for use in research (Chan et al. 2001).

OBJECTIONS TO GENETIC MODIFICATION OF ORGANISMS

Critics have raised a number of different objections to genetically modified organisms (GMOs), including GM crops, foods, plants, and animals. Some of these concerns are as follows:

GMOs Violate the Natural Order

For example, scientists have transferred genes across species and phyla to produce organisms that cannot be produced through natural breeding methods. Bioengineers have transferred antifreeze genes from fish into tomatoes to enable these fruits to have a longer shelf life. Bioengineers have also created mice that develop heart disease or cancer in six weeks and mammals that can only give birth via cesarian section. Many of the GM organisms contain human genes, a situation that many people regard to as repulsive.

The reply to this objection is that the mere fact that something is unnatural does not imply that it is immoral, since almost all modern science and technology is unnatural in some way. Non-GM crops, such as seedless watermelons and grapes, are unnatural, and non-GM animals, such as toy poodles and Siamese cats, are also unnatural. Unless we are prepared to reject modern science and technology, we should not find this argument against GM crops to be at all persuasive.

GMOs are Against God's Will

Many have argued that GMOs violate God's laws because God designed all natural species to perform specific functions in the world. Human attempts to redesign species or create new ones interfere with God's creation.

The reply to this objection is that selective breeding is also against God's will, if one holds that God intended for species to never evolve. Unless one wants to adopt the view that human beings should never interfere in natural selection and evolution, this view is also not very persuasive.

GMOs are Dangerous to Human Beings

Many critics object to GM foods and animals on the grounds that we do not yet have a firm understanding of their potential ill effects in human beings. GM foods could cause cancer, allergies, toxic reactions, and other problems, according to critics. They should not be considered safe for human consumption. Genetically engineered animals used to produce milk or organs also poses unknown risks to human health. Although most plant and animal hybrids do not threaten human health, critics argue that GM plants and animals create new and unknown risks not found in plant and animal hybrids, due to the anticipated effects of gene transference. In Europe, many people have opposed and continue to oppose GM foods. Although opposition to GM foods has not caught on as fast in the United States, many consumers are demanding that GM foods have a special label. Even African countries facing severe famines have been hesitant to accept GM crops because of these issues related to the safety of GMOs (McDowell 2002).

This objection to GMOs is different from the objection critics of genetic engineering voiced during the 1970s, when many people, including many prominent scientists, were concerned that genetic engineers would accidentally create a super bug that would escape the laboratory and infect and kill human beings. Scientists responded to these concerns during an historic conference at Asilomar, California in 1975 and recommended that researchers and the government regulate and oversee recombinant DNA experiments (Berg 1975). In the case of genetically engineered plants and animals, critics are not concerned about the risks of an accidental contamination or escape; they are concerned about the risks of intentional exposure to GMOs, since people are now eating GM crops and will soon be ingesting GM animal products (National Academy of Sciences 2002).

This is a potentially persuasive objection that requires further study. Although we have no evidence that GMOs pose a significant threat to human health and safety, we need more data concerning the risks that GMOs pose to human beings. We also need to develop mechanisms and processes for evaluating the safety of GMOs and for controlling their use.

GMOs are Dangerous to the Environment

According to this objection, GM crops can harm the environment in several ways. First, they can produce herbicides or pesticides that leach into the soil, enter the ecosystem, and threaten species. In one celebrated case, researchers discovered that genetically modified BT corn was killing monarch butterflies. Second, genes from GMOs may accidentally contaminate other species to produce superweeds that are resistant to herbicides. Third, some GMOs could disrupt native species if they leave the cultivated area and escape into the environment. A GM crop could become a superweed, like kudzu, and GM animal could be a superpest, like the fire ant.

This is another potentially persuasive objection that requires further study. Although we have no evidence that GMOs pose a significant threat to the environment, we need more data concerning these risks (National Academy of Sciences 2002).

GMOs Threaten Biodiversity

Critics have argued that by relying on GM crops instead of natural ones, farmers will eliminate biodiversity as they choose breeds for their productivity. For example, years of commercial breeding of corn led to a reduction in the diversity of the maize gene pool, which made these plants more susceptible to diseases. As GM crops continue to spread, natural biodiversity will be eliminated in agriculture. If GMOs contaminate the ecosystem, biodiversity can be lost outside of the agricultural realm as GMOs disrupt native species.

This is another potentially persuasive objection that requires further study. GMOs might have a negative impact on the biodiversity present in particular species, phyla, or ecologies, even if they do not have global effects on biodiversity. Although we have no evidence that GMOs pose a significant threat to biodiversity, we need more data concerning this risk.

Genetic Engineering Violates the Welfare of Animals

One might argue that many genetic engineering experiments violate the welfare of animals by causing suffering (NAS 2002). For example, researchers have designed GM animals to develop cancer, diabetes, obesity, hemophilia, Parkinson's disease, and many other human illnesses. One might argue that these modifications cause a great deal of animal suffering with little gain for human health or welfare.

This is also an important objection. To respond to it, it is important to remember that many animals suffered and died to meet human needs and interests long before the advent of genetic engineering. Most societies accept many types of animal experiments on the grounds that these experiments benefit human beings. Societies have also adopted a variety of animal welfare regulations and standards to protect the welfare of animals used in research (NAS 1996). These rules and regulations should apply to non-GM laboratory animals as well as to GM laboratory animals. If it would be unethical and/or illegal to use an animal in particularly cruel and pointless experiment, then it would also be unethical to design an animal for this express purpose. The larger question, which I cannot answer here, is whether it is ever morally acceptable to use animals in research (for further discussion see Lafollette and Shanks 1996). Unless one rejects animal experimentation per se, then the genetic engineering of animals for research purposes should be morally acceptable, provided that it pays appropriate attention to the welfare the animals and does not violate animal research rules and regulations. In any case, this situation bears watching, especially now that researchers are conducting experiments on primates, which demonstrate many significant cognitive traits, such as consciousness, memory, the capacity for language, and problem-solving skills.

Acceptance of GMOs can Lead Us Down a
Slippery Slope toward GM Human Beings

Once people become accustomed to GM plants and animals, they will be more willing to conduct gene transfer experiments on people to create the perfect human (Rifkin 1983, 1985, 1998). Not only would such experiments be extremely dangerous, but they would raise serious moral issues relating to eugenics, discrimination, human dignity, and social justice.

We encounter the slippery slope argument once again. The standard reply to this sort of worry is that one can stop the slide at some point by making careful distinctions, adopting appropriate regulations, and so on. Clearly, genetic engineering has come a long way since the 1970s. Whether we like it or not, the technologies required to genetically engineer a human being will soon be at our disposal. Thus, it is too late to stop part of the slide down the slope. Instead of trying to prevent people from developing these technologies for use on plants and animals, we should have thoughtful discussions about various topics related to genetic engineering, including cloning, embryo transfer, preimplantation genetic diagnosis, and gene therapy (somatic and germ-line), so that we can develop the appropriate policies and procedures to prevent the dangers of human genetic engineering, such as eugenics, the quest for perfection, discrimination, and so on (Resnik, Steinkraus, and Langer 1999).

Although I think it is worth mentioning objections to GMOs to provide some additional context for the patenting debate, It is not my aim in this book to explore all of these different critiques in detail.[1] Instead, I will focus only on those objections and concerns that relate directly to intellectual property issues. All of the objections listed above would arise even without patent and copyright protection, since the technology now exists to produce and design GMOs. Intellectual property rights encourage the development of these technologies and their applications in commercial areas, but the technologies required to make GMOs would have been discovered and used without any patent protection. Indeed, Chakrabarty developed his bacterium without any expectation that it would be patentable. Those who oppose GMOs on moral, religious, or philosophical grounds have opposed patenting as way of stopping the production of GMOs and the development of agricultural biotechnology, but patents are the wrong target, since researchers and companies would simply use trade secrecy to protect information if patents were not available. I now turn to those critiques that raise specific concerns about patenting per se, rather than general concerns about agricultural biotechnology.

THE LEGAL RELATIONSHIP
BETWEEN DNA PATENTS AND GMOS

Before addressing these critiques, I need to answer a prior conceptual question: what's the relationship between DNA patenting and GMOs? How could a DNA patent have any effect on GMOs? To answer these questions, recall that a patent grants the holder the right to exclude others from making, using, or commercializing his invention. This right includes a right to exclude others from using the patented invention in another invention, such as using patented parts in other inventions (*Telecomm Technical Services, Inc. v. Siemens Rolm Communications, Inc.* 2000). If someone uses a patented DNA part in a GMO, any attempt to make, use or commercialize that GMO could infringe the DNA patent. As far as patenting is concerned, a GMO is like any other invention that could be composed of different parts. Most of the parts in any given GMO are not patented, but some may be. For example, if someone isolated and purified a gene the produces herbicides in a plant, then they could patent that gene. A GMO containing the patented herbicide gene would infringe the DNA patent, unless the GMO's producer obtains a license from the DNA patent holder. Thus, DNA patents can have many downstream effects, including effects on proprietary rights in GMOs.

On the other hand, as noted earlier, the DNA patent holder's attempts to prevent other people from using his patented part could raise fairness issues under U.S. antitrust law (Lao 1999). If the patent holder is using his patent to wrongfully extend the scope of his patent to cover other products not addressed in his application, or he is attempting develop a monopoly over an entire market, or he is trying to prevent other people from entering the market, then it might be appropriate to use patent law and antitrust law to restrict the scope of his patent and to prevent him attempting to develop a monopoly.

Even though courts may refuse to recognize the DNA patent holder's attempts to exert control over GMOs that use his patented genes, someone who wants to develop and patent a GMO that contains a patented gene should probably negotiate a license with the patent holders in order to avoid a infringement lawsuit. In order to avoid seeking licenses from outside parties, a company could use genes that it had patented to develop a GMO. Thus, a biotech company making a new crop could maximize its proprietary control of the crop by patenting the plant itself, plants tissues, plant proteins, and plant genes.

With respect to GMOs, most of the ethical and political debate has focused on patents on the organisms themselves, rather than on patents on DNA, proteins, or cells contained in the organisms. Thus, most of my discussion in this chapter will focus on the patenting of GMOs, rather than on the patenting of DNA in GMOs. However, the points I will make in this chapter will apply to all types of patents related to GMOs.

ACCESS TO GM CROPS

One potential problem with patents pertaining to GM crops is that farmers in the developing world may have limited access to these crops, which may limit their ability to grow food for starving populations. To understand this problem, one must realize that for many years farmers in developing nations have collected and saved seeds to replant them for the following year's crops. Farmers in the developed world also engaged in this practice until the 1900s, but they have become accustomed to buying seeds every year from a seed company. Farmers in the developing world, however, often cannot afford to buy new seeds each year. Biotechnology companies have argued that farmers who save and replant such seeds from patented plants are infringing patent rights because the farmers have a license to use the plants, but not the right to reproduce or manufacture the plants from seeds. Indeed, some companies, such as Monsanto, have developed seeds that produce infertile plants in order to prevent farmers from replanting seeds. To grow these plants, a farmer must buy new seeds every year. This innovation completely circumvents natural reproduction in favor of artificial reproduction and makes the farmer dependent on the seed company for his or her livelihood, if he wants to grow the patented crops (Shiva 1996).

While *Chakrabarty* set the legal precedent for patenting organisms in U.S. law, *Ex Parte Hibberd et al.* (1985), provides the legal basis in US law for the patenting of GM plants. Kenneth Hibberd, a scientist working for the biotech firm, Molecular Genetic Research, Inc., had applied for a patent on genetically engineered corn. The patent examiners denied Hibberd's patent application on the grounds that Congress had intended plants to be patented only under the Plant Patent Act (PPA) or the Plant Variety Protection Act (PVPA). But the Appeals Board ruled in favor of Hibberd and followed the rationale for patents on living things established in *Chakrabarty* (Kevles 2001). Although Hibberd patented the GM plant, he could have also patented the DNA and proteins in the GM plant as well as techniques from making the GM plant. While this decision has no authority outside of the US and its territories, other countries have followed the precedent for patenting plants set in *Hibberd* (Kevles 2001).

Once someone has the legal right to patent a GM plant, how do they obtain the right to prevent others from producing the plant? As noted in chapter 3, a patent holder has exclusive rights to prevent others from making, using, reproducing, or commercializing his invention. The patent holder may assign some of these rights to others through a licensing agreement, and he may choose which rights he will assign. A patent holder might license someone to use an invention but not reproduce it or sell it. Since those who patent GM plants have the legal right to prevent other people from producing those plants without permission, the unauthorized growing of patented plants from seeds (or cuttings) could be construed as patent infringement. Indeed, companies that produce GM crops, such as Monsanto, require farmers to sign a licensing agreement that forbids them from replanting seeds. This is an important development in patent law because neither the PPA nor the PVPA prevented farmers from saving seeds to replant them. The PVPA has a farmer's exemption that allows farmers to save seeds for their own crops but not to save seeds in order to sell them to other farmers for planting (Plant Variety Protection Act 1970).

In evaluating the patenting of DNA used in making GM plants, my aim is to consider whether this practice poses a significant threat to agriculture. Since farmers in developed nations have adapted to buying seeds from seed companies and can afford to buy seeds each year, this practice does not appear to pose a major threat to agriculture in developed nations. The farmer who must buy his seeds every year is no worse off than the furniture maker who must buy wood, the automobile manufacturer who must buy steel, or the farmer who must buy water. Seeds are commodities that can be bought, sold, and used in manufacturing food. The free market will set a fair price for seeds and other agricultural commodities, provided that no single company obtains monopoly power and consumers have sufficient information and free choice.

On the other hand, the patenting of GM crops may pose a significant threat to agriculture in developing nations, since farmers in these countries often cannot afford to buy seeds each year. If these farmers cannot afford to buy seeds each year, then the world cannot obtain the full benefits of GM crops for world food production. If we take a utilitarian perspective on this issue, then we could say that we should make adjustments in patenting policies related to GM crops in order to reap the maximum social good. Policies that do not allow farmers in the developing world to reuse their seeds deprive society of important benefits. On the other hand, since patenting also yields important benefits for society and a failure to give adequate patent protection can result in harms to industry, one must carefully weigh and consider the consequences of honoring (or not honoring) plant patents in the developing world.

Although I am generally in favor of patents, even in the developing world, I can see how one might develop an argument for allowing farmers in the developing world to reuse seeds similar to public health exemption for drugs. The argument for the exemption in agriculture would take the same form as the one in public health: if a country is facing a famine of great proportions similar to the HIV epidemic in the sub-Saharan Africa, then it is justified in taking steps to restrict or override patents on agricultural products, such as GM crops. Many developing nations are currently facing severe famines, and have faced food shortages for years. In these countries only, farmers should be granted a "farmer's use" exemption similar to the exemption found in the PVPA. The exemption would allow farmers to grow plants from seeds for their own use but would not allow farmers to grow seeds in order to sell them to other farmers. Since this policy would still give biotech companies substantial control over their patented products, it would strike a fair balance between the demands of industry and the needs to alleviate severe famine in developing nations. Companies will not lose their entire market for seeds, since farmers in developed nations will still buy seeds, and farmers in developing nations will still buy new varieties.[2] If companies disagree with this policy, they have the option of offering farmers discounts for seeds to give them an incentive to buy seeds instead of replanting them.

As an aside, one might also argue that growing a plant from seeds should not be considered patent infringement at all because reproduction is a natural process, not a human invention. In chapter 3, I argued for distinguishing between products of nature and products of human ingenuity, based on the degree of human intervention required to create the item in question. If a thing would not have existed without human intervention, then we can treat it as a human invention. Without a doubt, GMOs would not exist without human intervention. Thus, GMOs should count as human inventions. But no human ingenuity is involved in natural (i.e., wild, uncultivated) reproduction by plants of animals, which occurs without human intervention. As long as human beings do not direct or control the reproductive process, it should be viewed as natural, even if it involves the natural reproduction of a GMO. If we take this perspective on the problem, then it follows that patent rights on GMOs should extend only to the cultivation and commercialization of GMOs, since these processes are both artificial processes. Patent rights may also extend to artificial reproduction of GMOs through cloning, cuttings, or other artificial methods, but patent rights should not extend to the natural reproduction of GMOs, otherwise the patent holder would gain control over something that he/she did not invent. Thus, a person who designs and patents a genetically engineered plant should be able to exclude others from cultivating, using or commercializing the plant, but he should not have the right to prevent the natural reproduction of the plant. Someone who purchases and grows the plant should not be held liable for patent infringement if his plants fertilize and reseed themselves, otherwise the patent holder would have a patent on a natural process. Thus, cultivating a patented plant from a seed or a cutting could constitute patent infringement. Even so, I would still argue that farmers in developing nations should be granted an exemption for this potential infringement, provided that they cultivate plants for food, not to sell seeds.

Some seed companies have decided to bypass this infringement problem by designing crops that cannot be propagated from seeds. In 1998, the U.S. Department of Agriculture and Delta & Pine Land Company were awarded a patent on a technique for manufacturing GM plants that produce infertile seeds. If the seeds are planted, they will not germinate. Because this technology is designed to terminate seed production, it is known as terminator technology (Shand 1999). Some critics, such as Shand (1999), have argued that terminator technology poses a threat to the world food security and biodiversity. She argues that companies should not be able to produce, distribute, or patent these GM crops.

Once again, this is a problem that will have its greatest impact on the developing world, since farmers in developed nations will be able to afford to buy seed each year. One might argue that terminator technology does not pose a significant threat to accessibility to GM crops because farmers in the developing world do not have to buy these seeds. They can buy seeds that can be replanted. Indeed, if farmers in the developing world do not buy these terminator crops, then the demand for these crops will weaken, which will encourage companies to develop crops that can be replanted, in order to make a profit in the developing world. Thus, one potential solution to terminator crops is to simply boycott those crops. On the other hand, a boycott might not be very effective if all companies develop this technology and it becomes an industry standard (Shand 1999). In this case, the market would not be able to provide farmers with nonterminator seeds and farmers would have no choice but to buy terminator seeds.

Although I think the problem of terminator crops is a plausible threat, it does not currently pose a problem for access to crops in the developing world, because farmers can buy nonterminator crops. However, countries should continue to monitor this situation and they should be prepared to take steps to prevent monopolistic restrictions on seed commerce, to insure that non-terminator crops are always available. Another reasonable precaution would be to closely monitor this technology to ensure that terminator genes do not escape and infect other plants. Companies should be required to demonstrate that the risks of accidental contamination are exceedingly low and they should develop methods of minimizing the damage from terminator genes if they are transferred to other plants. For example, one could attach a termination gene to the terminator gene, which would deactivate the terminator gene in response to a chemical signal.

Thus, the problem with access to GM crops in the developing world is a plausible threat posed by patenting in agriculture, but the threat is not sufficiently dangerous to justify a ban patents, including DNA patents, in agriculture. Instead, countries should take steps to insure that farmers in developing nations have access to GM crops, such as adopting a farmer's use exemption for GM crops and regulating the market for terminator crops.

EXPLOITATION

Another potential threat to agriculture brought about by the patenting of DNA is the exploitation of farmers and native peoples. In Chapter 8, we addressed the topic of exploitation as it relates to the doctor-patient relationship. The general approach I took in that chapter was to view exploitation as taking unfair advantage of someone. I also discussed several different ways that exploitation can occur. Shiva (1996) argues that patents allow biotech companies to exploit poor farmers in the developing world. She says that the intellectual property rights asserted against developing nations conform to the pattern of exploitation embodied by colonialism. In Shiva's view, biotech companies from the industrial world are taking unfair advantage of people in the developing world by selling them seeds and by developing patented products from their traditional knowledge of plant medicines (Shiva 1996).

How might companies be taking advantage of people in the developing world? First, one might argue that they are engaging in economic transactions that violate informed consent because these deals result from fraud, deception, coercion, or undue influence. Second, one might argue that they are taking advantage of desperate circumstances to induce unfair bargains. Third, one might argue that the deals companies make with people in developing nations are unjust even if they are not induced unfairly. Let's focus on the problem of seed selling first. Does this violate the informed farmers in the developing world? Probably not, since it is likely that farmers that buy seeds make informed choices. Barring some evidence that these deals result from fraud, deception, coercion, or some other problem with the consent process, the deals probably represent the informed choices of farmers. Should we regard such deals as unfair inducement, due to the desperate circumstances of the farmers? Again, the answer is probably no, since in most cases the farmers have the option of buying non-GM seeds. Even though these seeds are not as productive as GM seeds, they present farmers with a viable option as long as they are available.

Thus, the exploitation problem concerning seed selling boils down to an issue concerning the injustice of selling GM seeds in the developing world: are farmers getting an unfair or inequitable deal? Few people would argue that farmers in the developed world get an unfair deal from seed companies, since these farmers can afford to pay the market price for seed. So what is different about the developing world? Could a deal be fair in Kansas but not in the Sudan? This is a difficult question to answer, since it asks us to compare farmers in two very different situations. However, one might argue that farmers in the developing world are getting an unfair deal from seed companies because the deal prevents them from obtaining what is a very important benefit for them (i.e., the right to replant seeds). This right is also denied to farmers in Kansas, of course, but these farmers do not value this right as much as they value productivity and convenience. So, I would say that the current practice of not allowing farmers in the developing world to replant GM crops is unfair and exploitative. This situation can be avoided, however, if we make some changes in patent law and policy, such as the farmer's exemption (discussed above), which are designed to give these farmers a better deal.

The other exploitation problem mentioned by Shiva (1996) concerns the patenting of products derived from traditional knowledge, sometimes referred to as bio-prospecting (Rifkin 1998). Many indigenous cultures have a great deal of knowledge about herbs and medicinal plants that they have gained from their experiences with those plants, which they have passed down for hundreds of generations. Often, this knowledge is not written down but is transmitted orally from shaman to student. In many cases, researchers have patented products, such as pharmaceuticals, tissues, and genes, developed from this traditional knowledge. Large pharmaceutical companies, such as Eli Lilly and Merck, have become interested in developing drugs from the knowledge of native healers. For instance, Merck paid $1 million to INBio, the National Biodiversity Institute of Costa Rica, for rights to develop drugs from plant samples gathered in the Costa Rica's rain forests. In another deal, Eli Lilly paid $4 million to Shaman Pharmaceutical for bioprospecting rights. The deal required an unspecified amount of money to be given to the Healing Conservancy, a nonprofit corporation that promotes conservation (Shiva 1996). According to Shiva, these are unfair deals for the developing nations, especially when one considers the potentially huge market for pharmaceuticals derived from traditional knowledge of native plants. Indeed, these deals are similar to exploitative agreements between the developed world and Native Americans for land, Africans for diamonds, or Arabs for oil.

Are these deals exploitative? Let's first consider the issue of informed consent. If the company does not offer anyone a deal but simply steals the knowledge, then this would be clearly exploitative. An improperly negotiated deal could also violate informed consent. For instance, if a pharmaceutical company strikes a deal with a government in a developing nation, and the government does not consult with or get approval from the people, then one might argue that such a deal is exploitative because it violates the informed consent of the people in that country. Thus, one of the most important conditions that should be observed in these deals is the duty to inform the people who will be affected by the deal and to obtain their consent. Next, we should consider the issue of undue influence: any deal between a company and people in a developing nation that takes advantage of their desperate circumstances to induce a bargain would result in undue influence. Since the people in these nations have reasonable alternatives to striking deals with companies, undue influence is probably not a major concern in these transactions.

Now we come to the final issue, justice. Are these deals equitable? Are these people getting their money's worth when they sell rights to use their traditional knowledge of herbs and medicinal plants? To answer this question, we really need to look at the details of the agreements that have been reached. A lot depends on the market value of the rights that are sold, which is hard to determine, given the tremendous risks of drug development. Shiva (1996) thinks that the rights with a potential market value of billions of dollars have been sold unfairly for millions of dollars. Clearly, any deal that exchanges something worth a million dollars for something worth a billion dollars is inequitable, and a court of law might even find such a deal to be unconscionable (Rohwer and Skrocki 2000). But once again, we return to the key empirical question: what is the market value of the item that is allegedly the subject of an unfair transaction? Bioprospecting is prospecting. As long as the risk to industry remain high, it will be difficult to show that deals that are struck between companies and native populations are unfair. Perhaps a partial solution to this dilemma would be for communities to ask for a share of royalties from products developed from their knowledge in addition to a fee for using the knowledge. Royalties would give the community a share of the profits if it turns out that a company is able to develop a blockbuster drug.

If we move beyond individual cases of exploitation, such as deals between pharmaceutical companies and countries, do we have any evidence that DNA patents in agriculture are contributing to some general trend of exploitation? In other words, do DNA patents make the relationship between the developed world and the developing world more or less exploitative? This is a very difficult question to answer as well. To get some kind of a handle on this question, one would need to summarize all the economic transactions between the developed world and the developing world to determine whether there is an unfair imbalance of benefits and burdens. Is the developing world taking a great deal from the developing world and providing very little in return? On the one hand, one might argue that developing nations are taking natural and cultural resources from developing nations and providing only money in return. While this is less exploitative than outright theft or conversion, it is not terribly fair. On the other hand, one might argue that developing nations not only obtain money but also obtain education, infrastructure, technology, and economic development.

An even more difficult question to answer is whether the intellectual property system in general makes the relationship between developing and developed nations more or less fair. Shiva (1996) argues that intellectual property rights encourage the transfer of resources from developing nations to developed nations for marginal compensation. Without a doubt, developed countries have benefited a great deal from the intellectual property system. Developed nations own most of the patents, copyrights, and trademarks in the world, they fund most of the world's research and development, they create most of the world's knowledge and technology, and so. But have developed countries prospered at the expense of developing countries? Have they gotten rich by robbing resources from developing nations?

To answer these questions, it is important to remember that intellectual property is different from tangible property in two respects that are relevant to exploitation issues. First, intellectual property can be shared; it is naturally non-exclusive. Thus, there is not a limited supply of intellectual property. To create an invention, one doe not need to take it from someone else. Second, intellectual property does not exist in any particular place, such as a particular country or town. Obtaining knowledge from the developing world is therefore very different from obtaining natural resources, such as coal, oil, gems, fur, or bones from the developing world. If one extracts oil from a country, the oil is gone and people can no longer use it. But if one acquires knowledge from a country, the people still retain their knowledge. The oil also has a definite location that justifies property claims to the oil: the oil is under their land, so it belongs to the people who possess that land. Knowledge, on the hand, does not have a definite location. Although people in particular communities acquire and develop knowledge, once the knowledge is disseminated, it is no longer located within that community. The Ancient Greeks developed geometry, but geometry no longer belongs to the Greeks. Indeed, the whole world has benefited from Greek geometry.

Thus, intellectual property transactions do not involve taking something from one party and giving it to another party. One does not gain something at another's expense. Instead, both parties can benefit. However, even if both parties can benefit from an exchange of intellectual property, one might argue that the exchange can still be unfair if one party benefits much more than the other party. If A transfers his copyrights to his song to B for $10,000, and B makes $10 million from that song, we would say that this transaction was unfair, even though both parties have benefited from it. So we are back to the question we asked earlier: when developed nations have acquired knowledge from developing nations, have these transactions been fair exchanges? I am not sure that we have a clear answer to this question yet. Although we have good reasons to believe that intellectual property benefits everyone in society (including all people in the world) by helping to promote the growth of science and technology, questions concerning the overall (or global) fairness of the intellectual property system merit further study. We do not have enough evidence at this point to say with any confidence that intellectual property rights contribute to or do not contribute to the exploitation of the developing world.

Having said this much, I would still admit that the exploitation of developing nations is a plausible threat posed by DNA patents; it is not some outlandish nightmare or arcane idea. Exploitation is a genuine threat that needs to be taken seriously. We should take precautionary measures to prevent exploitation as well as proactive measures to promote fairness. For example, all deals between pharmaceutical and biotech companies and developing nations should be carefully scrutinized to monitor and assess concerns relating to autonomy, undue influence, and fairness. Companies should follow duties of corporate responsibility in interacting with developing nations, and they should help these countries develop their education, infrastructure, technology, and economic systems. All agreements pertaining to global trade and intellectual property, such as TRIPS and GATT, should also be evaluated with an eye to justice and equity to insure that developing nations get a fair deal in international treaties, transactions, and negotiations.

THE COMMODIFICATION OF AGRICULTURE

The final threat to agriculture posed by DNA patents is the familiar issue of commodification. Shiva (1996), Rifkin (1998), and others have argued that the intellectual property scheme is contributing to the commodification of food, crops, and farming. Food is becoming a commodity and agriculture is becoming an industry. Today's world of large commercial farms, fertilizers, pesticides, herbicides, and farm machinery is a far cry from the agrarian ideal of natural, small-scale, family farming. Recalling the distinction between complete and incomplete commodities, this argument asserts that intellectual property rights are transforming food into a complete commodity and agriculture into a completely commercial activity.

While I am sympathetic to this critique, noncommercial agriculture is a romantic ideal that is fast fading into the annals of history. Although there is something attractive about a small family farm, commercial farms are far more efficient at producing food. Agrarian societies, which have existed for the last 10,000 years of human history, are being replaced by modern, industrial societies. Given the enormous task of feeding the world's burgeoning population, noncommercial food production must give way to commercial food production. It may still be possible to preserve some of this romantic, agrarian ideal in various places around the world, much in the way that the Amish resist modern technology, but from now on it will be the exception rather than the rule and a curiosity rather than a standard mode of operating. Thus, while one may curse the effects of intellectual property on agriculture, and mourn the demise of the agrarian way of life, these considerations have very little import for our policies on biotechnology patents. The transformation of agriculture began during the scientific revolution, long before countries recognized patents on living organisms (Volti 1995; Jacob 1997). From the rotation of crops, to the selective breeding of hybrids, to the use of fertilizers and pesticides, agriculture has become increasingly industrialized and commercialized. A ban on biotechnology patents might slow down the commercialization of agriculture and impede food production, but it would have no substantial effect on this historical and economic trend toward increasing industrialization. Thus, commodification is not a threat to agriculture because it has already happened. It is pointless to take precautions against an event that has already occurred, and imprudent to try to reverse a trend that has the weight of the world's hungry masses behind it.

SUMMARY

This chapter has considered the potential threats to agriculture posed by DNA patents and argued that patents pose some very real threats to the accessibility of GM crops in the developing world and raise potential exploitation concerns. In order to ensure that developing nations are able to obtain the full benefits of agricultural biotechnology, farmers in the developing world should be granted a farmer's exemption to patent infringement to allow them to replant seeds that they have saved from biotech crops. Replanting should be allowed because it constitutes a humanitarian response to the problem of world hunger. In order to avoid exploitation, the bio-prospecting activities of biotech and pharmaceutical companies should be carefully monitored and scrutinized to ensure that private corporations make fair and equitable deals with native populations. Communities that have traditional knowledge about herbs and medicinal plants should receive a fair price for the use of that knowledge by researchers and private firms.

10

Conclusions and
Policy Recommendations

SUMMARY OF THE BOOK

In this book, I have attempted to provide the reader with an overview of the moral and policy issues related to the patenting of DNA. The first two chapters covered the scientific, technological, and legal aspects of the topic. Subsequent chapters discussed a variety of moral arguments for and against the patenting of DNA products and processes, and made a number of suggestions relating to public policy. It is now time to take stock of these arguments and summarize these recommendations.

I have classified moral arguments for and against DNA patents as either consequentialist or deontological. I have argued that deontological arguments play a minor role in the case for or against DNA patenting. Deontological arguments do not play a significant role in arguments for DNA patents, because there is no substantive moral or legal right to patent: a patent is a privilege granted by the government to promote social goals, such as the progress of science and the practical arts. Thus, the government can (and should) restrict patents in order to promote social goals. Deontological arguments do not play a significant role in arguments against DNA patents because very few DNA patents would be inherently immoral. Patents on a whole human genome, a human embryo, or a human being would be inherently immoral because they would violate human dignity. Any patent on a single gene that was interpeted as applying to a whole genome, embryo, or human being would also violate human dignity. These types of patents should be regarded as unethical because they treat people, developing people, or human body parts intimately related to personal identity, as mere commodities or things. Patents that violate human dignity should be illegal.

Since deontological considerations related to moral rights and duties play a minor role in arguments for or against DNA patenting, most of the moral analysis depends on a careful examination of the consequences of DNA patenting for science, medicine, agriculture, industry, and society, and a judicious balancing of the various interests and values at stake in this debate. This book has examined some of these potential consequences in chapters 7 through 9, which dealt with the impact of DNA patents on science, medicine, and agriculture, and in chapter 6, which addressed moral concerns related to the commodification (or commercialization) of human body parts.

If we could assign objective probabilities to these potential consequences, and we had a procedure for ranking our values and interests, then we could balance these interests and values by adopting rules and policies designed to maximize expected utilities.[1] Under these ideal conditions, we could adopt a cost-benefit or risk assessment approach to DNA patenting. However, there is still a great deal we do not know about the consequences of DNA patenting for society. We have a pretty good idea of some of the general benefits of patenting: over 400 years of historical evidence demonstrate the patents stimulate scientific discovery and technical innovation, which lead to practical applications in medicine, agriculture, engineering, and industry. We also know some of the potential harms the result from abuses of the patent system, such as anticompetitive practices and underutilization of scientific and technical information. Although we have a sound understanding of the potential benefits of DNA patenting, we do not yet have enough evidence to assess the potential harms of patenting. At this point in time, we can only say that DNA patents will probably produce many benefits for science and society, but that they may also cause various harms. Since we cannot assign objective probabilities to the potential harms of DNA patents, we cannot apply decision strategies that seek to maximize expected utility, because we cannot assign objective probabilities to all of the outcomes we need to calculate expected utilities.[2] Many of our decisions concerning patenting policy should be regarded as decisions under ignorance, rather than decisions under risk. If we had sufficient time and resources, we could gather more evidence in order to reduce our uncertainty and make accurate predictions about the consequences of various policies. However, we do not have the luxury of waiting of waiting for more evidence, and we must make some decisions concerning patenting policy, despite our high degree of uncertainty. Even a decision to do nothing is a decision.

I have argued that the Precautionary Principle can be a useful strategy for making decisions relating to patenting policy when we lack enough evidence to implement a strategy designed to maximize expected utilities. In order to apply this principle to DNA patenting policy, we should take reasonable precautions to avoid plausible threats to science, medicine, agriculture, and society. We can decide whether a threat is plausible even when we lack enough evidence to assign an objective probability to the threat. Some of the threats this book has considered include:

- Threats to progress, openness, and integrity in science;
- Threats to the practice of medicine, including exploitation, commercialism, conflicts of interest, and problems with access to care;
- Threats related to agriculture, including exploitation, commercialism, and problems with access to genetically modified crops.
- Threats to society, including threats to human dignity, exploitation, and a loss of respect for the value of human life.

Given these threats posed by DNA patents, we have three basic social choices:

1. Ban DNA patents.
2. Do nothing; maintain the status quo.
3. Develop policies, rules, and regulations designed to prevent these threats or minimize their impact.

I think option 1 would an unreasonable response to the moral, social, and economic problems related to DNA patents, since patents are likely to yield many benefits for society. In science, DNA patents stimulate discovery, innovation, and private investment in research; in medicine, they encourage the development of new tests, new procedures, and treatments; in agriculture, DNA patents encourage the production of useful crops, plants, and animals, as well as private investment in agricultural biotechnology. A ban on DNA patents would be an overreaction to the threats they pose; it would be throwing the proverbial baby out with the bath water. On the other hand, sitting still and doing nothing, Option 2, would also be unreasonable, because the threats are real and could have profound effects on science, medicine, agriculture, and society. Option 3 is the most reasonable course of action. We should continue to allow DNA patenting and should develop policies designed to prevent these threats or minimize their impact. The next section summarizes some of those policies.

POLICY RECOMMENDATIONS

A brief summary of the patenting policies discussed in this book is as follows:

1. Patents rights should not be construed as granting patent holders exclusive control over naturally occurring DNA sequences or naturally occurring

genomic processes. Patent holders may be granted exclusive control over artificial sequences (e.g., brand new DNA sequences or isolated and purified sequences). Patent holders may also be granted exclusive control over methods or techniques that use natural processes to achieve specific results. Anyone should be able to use, make, or commercialize a naturally occurring DNA sequence without infringing on patented versions of the sequence. Likewise, anyone should be able to use, make, or commercialize a naturally occurring genomic process without infringing on patented inventions that use that process. Naturally occurring DNA sequences and genomic processes should remain in the public domain. If a patent is interpreted as granting the patent holder exclusive control over naturally occurring DNA, then it should be rejected as excessively broad in scope.

2. There should be no patents on a whole human genome, a human embryo, or human being. Patents on parts of the genome (e.g., genes and gene fragments) are acceptable. However, a patent on a part of a genome should not be interpreted as giving the patent holder proprietary control over the whole genome, a human embryo, or a human being.

3. Patenting agencies and the courts should carefully define the scope of DNA patents in order to provide incentives that stimulate invention and investment but do not undermine access to technology or entry into the free market. DNA patents should not be so broad that they stifle competition and give a patent holder monopoly power of a wide range of markets. As noted above, a DNA patent should also not be so broad that it grants the patent holder control over naturally occurring DNA.

4. Patent agencies should set a fairly high standard for proving the utility of a DNA sequence. Applicants must provide substantial proof that the sequence described in the application has potential practical uses in diagnosis, treatment, prevention, drug discovery, or genetic engineering. Researchers should not be allowed to patent DNA sequences based on speculative or whimsical claims pertaining to usefulness. Asserting that a particular gene carries information to make a specific protein may establish the biological function of the gene but it does not, by itself, prove the gene has a practical, human use. To establish the usefulness of the gene, one must also show how one can use its protein product for some practical purpose, such as drug development or diagnosis. In order to patent an EST, a researcher must be able identify its corresponding gene and demonstrate the usefulness of that gene.

5. Patent agencies should also keep close tabs on the obviousness requirement to ensure that DNA inventions represent genuine advances beyond the prior art that would not be obvious to those trained in that art. The fact that DNA sequencing has become much easier, due to automation,

does not render all sequences produced by automation obvious. To be nonobvious, an invention need not result from a flash of genius or from a greater deal of labor; it must only represent an invention that is not described or used in the prior art.

6. Patent agencies should consider taking some steps to ease the pressures to maintain secrecy prior to filing a patent application. One way to do this would be for patent agencies to encourage researchers to file provisional patent applications and to assure them that publication of data related to their invention will not invalidate their patent, as long as they do not publish enough data to allow someone trained in the prior art to make and use the invention. It would also be helpful to extend the provisional patent period from one year to three or more years. Extending the length of a provisional patent would require legislators to revise the patent laws, however.

7. The courts, patent agencies, and researchers should make better use of the research (or experimental use) exemption in patent law, which allows researchers to use or make patented inventions for noncommercial purposes, such as basic research. However, since the line between commercial and noncommercial research is becoming increasingly difficult to draw, the legal system should take great care to define this exemption, since the exemption should not apply to academic researchers who are conducting research that may have commercial applications. Although there is a basis for the exemption in federal common law, additional legislation may required to articulate, specify, and justify this exemption (Nuffield Bioethics Council 2002).

8. Likewise, farmers in the developing world should be granted a farmer's exemption to allow them to use patented seeds and tissues or plants that contain patented genes. These farmers should be allowed to save and replant patented seeds for personal use without fear of patent infringement, but they should not be allowed to save seeds in order to sell them to other farmers. To implement his exemption, countries with an interest in this issue will need to reach international agreements related to a farmer's exemption, and developing nations will need to pass new legislation that describes this exemption. It is also important to define this exemption clearly, and to avoid ambiguities relating to important terms, such as farmer and developing world.

9. Private companies and public institutions should develop databases for DNA, RNA, proteins, and other important information in biotechnology. Copyright protection should be extended to these databases, provided that they meet the originality requirements established in copyright law. The doctrine of fair use in copyright law should apply to these databases. Thus,

noncommercial users should be allowed to copy small parts of these data-bases for personal use, provided that such copying does not defray the commercial value of the database. Large copying of databases for public or private purposes should not be allowed. Copyright laws should not protect any of the facts in these databases, which should remain in the public domain.

10. Public and private sectors of the research community should continue to cooperate in sharing data, materials, and tools. They should cooperate on large projects, such as genome sequencing, distributed computing, and information consortiums.

11. Basic researchers as well as clinical researchers should disclose potential conflicts of interest to concerned parties including journals, employers, professional societies, administrators, supervisors, students, and patients. In some situations, these conflicts should be managed carefully or should be prohibited completely. Institutions should also attend to their own conflicts of interest arising from patenting and commercial activities.

12. When researchers collect cell or tissue samples from subjects, they should inform the subjects of the potential economic value of their cells or tis-sues (Clayton et al. 1995). This disclosure should also mention the eco-nomic value of any potential DNA patents. Although researchers do not have to offer subjects any financial compensation for their cells or tissues, they should inform subjects of their plans to offer (or not offer) compen-sation. Subjects may agree to forego compensation only if their decision to waive potential proprietary rights is free and informed.

13. Companies that develop tests and treatments from patented DNA should address issues related to benefit-sharing in genetics research (Merz et al. 2002). They should consider offering financial or other compensation to tissue donors or communities affected by genetic diseases in order to avoid exploiting patients and their families.

14. Governments should monitor bioprospecting deals between pharmaceu-tical or biotech companies and local populations and should not approve of unfair deals. Local populations should seek agreements that grant them a share of royalties in addition to an upfront fee.

15. In order to address ethical issues in biotechnology patenting, including DNA patenting, patent agencies should consider developing some form of ethical review for patents (Gold and Caulfield 2002). An ethics review committee would not attempt to assess the scientific or legal aspects of patents; it would only address concerns related to an invention's ethical aspects. In order to avoid unnecessary impediments in the application process, the committee should only review a small set of applications

that have significant ethical questions. Patent examiners should be granted some discretion to refer patents for ethics review. If the committee determines that an invention would violate or significantly threaten a fundamental ethical value, such as respect for human dignity, then it could recommend that the agency delay issuing the patent, pending further review. For example, a patent review committee would be able to offer an ethical assessment of a patent on a human embryo or a "humanzee." The committee would not have the power to block patents; its opinions would only be advisory. Nevertheless, an ethical review committee could play an important role in addressing potential ethical issues in patenting, and it could serve as a liaison between the patent office and the public (Gold and Caulfield 2002). A patent review committee would be similar to other bioethics committees that have given advice to the executive branch of government, such as the President's Council on Bioethics, the National Bioethics Advisory Commission, and the Recombinant DNA Advisory Committee.

16. There is no need, at the present time, to make any fundamental changes to the patent system or enact major legislation to deal with the ethical, social, and political issues relating to DNA patents. Governments may need to adopt some minor legislation to define the two exemptions to patent law defended in this book, the research exemption and the farmer's exemption, and to extend the length of provisional patents. Countries may also need to reach some international agreements concerning the farmer's exemption. The existing legal and regulatory framework, including patent laws, copyright laws, antitrust laws, human research protection rules, the courts, patent agencies, and human research agencies can handle almost all of the problems that currently exist as well as problems that are likely to arise in the near future. To preclude the need for additional regulation, regulators, administrators, and judges should make effective use of the existing laws and rules in order to develop effective patent policies, which should respond to the concerns and recommendations discussed above. If economic and technological conditions change, there may be a need to revise the patent or copyright laws in the future, but at the present time, making any substantial changes in the laws could do more harm than good.

This book is not the first word written about DNA patenting and it won't be the last. It has been my hope only to advance public and scholarly debate about DNA patenting, not to solve all of the problems once and for all. DNA patenting raises important moral and policies issues affecting many different sectors of the economy and society. As we begin the twenty-first century, we should be open to the exciting opportunities presented by industrial biotechnology but we should also take steps to safeguard human values. Patents on biological materials, such as DNA, provide crucial incentives for this developing industry, but they also threaten the progress of science, the practice of medicine, the development of agriculture, and the preservation of cultural notions concerning the value of human life. Since the future is uncertain, we should take great care to observe and assess developments in industrial biotechnology and take precautionary measures to promote an intellectual property scheme that provides incentives to inventors and entrepreneurs but also protects human values.

Notes

CHAPTER 3

1. Rights are moral, legal, or political claims that members of society can make against each other (Feinberg 1973). A person who has a right to due process in a legal system can make legal claims against people and governments who attempt to deny him of that right. A person who has a right to vote can make political claims against people and governments who would deny him this right.

2. Although I am careful to distinguish between property rights and the thing that is controlled by property rights I will frequently resort to the more common way of talking about property, which does not attend to this distinction. I do this to make the book more readable, not necessarily more clear or precise. Clarity sometimes must succumb to the demands of style.

3. I am assuming here that DNA sequences exist independently of human beings and are not original works. If one took an antirealist approach to DNA, it might be possible to argue that a DNA sequence can be copyrighted (but not patented) because it is an original work! On the realist approach, DNA exists independently of human beings and is not an original work, although a paper or other publication representing DNA would be an original work.

4. We will explore these slippery slope concerns in more detail in chapter 6.

5. This information was obtained by searching the PTO's database and using the general search terms, DNA, gene, cell, protein, and mouse. Readers should note that I am construing the category of patents related to a subject matter as broader than the category of patents on a subject matter. Thus, it is not self-contradictory to claim that the PTO has awarded about 16,000 patents on DNA but over 50,000 patents related to DNA.

6. As a side note to the DNA patenting debate, the reader should be aware that protein patents will become increasingly important as researchers develop applications of proteins in drug discovery and diagnosis. Indeed, protein patents may prove to be nearly as important in biotechnology as DNA patents, since human beings have as many as two million proteins but only 30,000–40,000 genes (Service 2001).

7. Reasonable and reasonably are also great wiggle-words in the law.

CHAPTER 4

1. The word *deontological* comes from Greek words *deon* (duty) and *logos* (law or science). Translated literally it means "law of duty." Since all moral theories imply moral duties, one could say that all moral theories, including utilitarianism, are deontological. But this is not how the word has been used in moral philosophy. According to some writers, deontological, refers to acts or rules that are morally obligatory regardless of their consequences (Olson 1967). As Rawls (1971, 30) notes, this view does not make a great deal of sense because "All ethical doctrines worthy of our attention take consequences into account in judging rightness. One which did not would simply be irrational, crazy." The better way of interpreting the word deontological is to say that deontological theories are not consequentialist, meaning that while such theories consider consequences they do not define moral rightness (i.e., right and wrong) via a direct link to consequences. Utilitarianism, on the other hand, makes a clear and direct connection between the definition of moral rightness and consequences.

2. For further discussion of controversies about rights, see Lyons (1979), Glendon (1991).

3. I use the phrase modern societies because some societies, including the US during the Colonial period, made property a prerequisite for basic rights under the law. Rights granted under the Constitution originally applied to adult, male, property holders.

4. There are some exceptions to my general claims that intellectual properties are not closely connected to individual welfare, dignity, and autonomy. For example, poems, songs, paintings, plays, and other artistic and literary works often bear a close and intimate connection to the individual author. These types of intellectual property would therefore merit greater protection than other types, on my view. This idea accords with Hegel's self-expression approach to property discussed in chapter 3: intellectual properties that are essential to self expression deserves greater protection than those that are not essential to self-expression.

CHAPTER 5

1. There is an important philosophical issue here about deciding what counts as human DNA. As stated earlier, only 2 percent of the DNA found in human beings occurs only in the species Homo sapiens, but this number is a bit of an oversimplification since those unique sequences are themselves composed of nucleotide base-pairs that are found in other species. An analysis of the human genome also reveals that we have a great deal of viral DNA in our genome due to infections from retroviruses over the course of human history (Subramanian et al. 2001). A retrovirus incorporates its DNA into the host's genome when it invades the host. Finally, as a result of biotechnology and genetic engineering, many human genes have been transferred into other species. For example, scientists have created pigs with DNA that expresses human antigens on cell surfaces. I will not provide a fuller analysis of the problem here, but I

would like to point out this interesting issue. Elsewhere I have argued for a functional, as opposed to a structural, definition of human DNA: human DNA is DNA that functions in human beings (Resnik 1997). Thus, even if a Ford truck shares parts with many other vehicles, a part is a Ford truck part insofar as it functions (or performs a role) in a Ford truck.

2. Many Native American tribes believe that people do not own nature, including animals, plants, and the land. One could adopt this type of view to argue against DNA patenting, but there would still remain the issue of whether the item in question is a natural thing or a human artifact.

3. There are, of course, many different ways of analyzing the word cause. Philosophers have analyzed causes in terms of necessary and sufficient conditions, while statisticians have analyzed causes in terms of statistical connections (Pearl 2000). Legal scholars distinguish between the cause-in-fact and the legal cause (or proximate cause). A cause-in-fact is a sine qua non condition that is roughly equivalent to a necessary condition. If John hit Jim with a baseball bat and Jim developed a concussion as a result, we can say that John caused Jim's concussion because but for his actions, Jim would not have had his concussions. In philosophy, one would say that John's actions were a necessary condition of Jim's injury. A legal cause has no relation to any physical notion of causality but is understood in terms of legal duties related to foreseeability, proximity in time and space, and moral responsibility (*Black's Law Dictionary* 1999).

4. By objective I mean independent of human wants, goals, theories, beliefs, biases, etc. When most people think of objective features of the world, they think about scientific facts about the world, such as physical, chemical, or biological phenomena. If one holds that some normative principles and obligations are true (or correct) independent from human wants, goals, theories, beliefs, and biases, then it would be possible for moral, political, and others norms to be objective as well. See Gibbard (1990) for further discussion of normative objectivity and Kosso (1992) for further discussion of scientific objectivity.

5. Some utilitarians defend a monistic theory of value and argue that all of these values can be understood in terms of one single value, such as happiness or the satisfaction of preferences. Other utilitarians accept a pluralistic theory of value and argue that one should maximize a number of different, incompatible values, such as happiness, justice, human rights, knowledge, and so on. I hold a pluralistic theory of value but I will not defend it here. For more on pluralism, see Rawls (1993), and Gutman and Thompson (1996).

CHAPTER 6

1. There are at least two different ways of distinguishing between intrinsic and extrinsic values. According to the definition I have given, a value (intrinsic or extrinsic) is a property of the thing itself and the words are treated as adjectives that modify nouns. Thus, an intrinsically valuable painting is like a red or eighteenth century painting. On the other hand, one might argue that a value is not a property of the thing

itself but is a property of the valuers. The distinction between intrinsic and extrinsic values reflects different modes valuation. A mode of valuation could be understood as a three-place relation involving a thing, a valuer, and a type of value. Thus, to speak of something as intrinsically valuable is really a way of saying to value intrinsically, for example, "I value the painting intrinsically." The words intrinsically and extrinsically, according to this view, are adverbs that modify verbs. Thus, valuing intrinsically is like seeing bluely, hearing imperfectly, or judging poorly. I think the former view is preferable because it is more in accordance with ordinary language, and it better reflects our experience of value. Defenders of the other view would point out that treating the words as adjectives that refer to properties of things commits one to a realistic approach to values because these values are understood to be properties of objects, not properties or functions of human cognition, judgment, or emotion. My response to this objection is that ordinary language and moral phenomenology do indeed assume a realistic approach to values. If we take ordinary language and experience at face value, then the burden of proof is on the critics to show why our language and experience rest on a mistake or illusion, and why we should adopt the revisionist approach to values. For further discussion, see Mackie (1977).

2. There is a dispute among Kantians about whether or under what conditions one would be morally justified in killing innocent people in order to save other people. According to some Kantians, it is sometimes acceptable to kill innocent people in order to save human lives provided that the reason for doing so is not to maximize overall good. Suppose a Kantian faces this situation: if we shoot down a commercial airplane we will kill 250 innocent people but save more than 4,000 human lives. A utilitarian would have no difficulty in carrying out this calculation and deciding that we should shoot down the airplane. A Kantian might reach the same result by a different path. For example, the Kantian might appeal to the universal law form of the CI and judge that the maxim "I will shoot down a commercial airplane with 250 people on board in order to save more than 4,000 lives," would be a rule that could be universalized in that it is conceptually consist and would be adopted by rational agents. A great deal depends on how one formulates the maxim (or moral rule) that serves as a basis for action. For further discussion, see Hill (1992), and O'Neill (1996).

3. I recognize that some people do not agree with capitalism and view it as morally bankrupt and politically exploitative. Marxists, for example, object to capitalism on the grounds that it exploits the poor for the benefit of the rich (Marx [1867] 1996; Marx and Engels [1848] 1998). There are two classic defenses of capitalism, the libertarian view and the utilitarian view. Libertarians, such as Nozick (1974) defend capitalism on the basis of natural rights to property: free markets allow people to express their values and trade their properties. Governments should not interfere with the free market unless regulations are required to prevent people from violating each other's rights. For example, a libertarian would allow laws designed to ensure fair dealing in the free market, such as laws against fraud, duress, theft, conversion, and the like. Though not known as a utilitarian, Adam Smith ([1776] 1991) provided a utilitarian argument for free markets in his notion of the invisible hand operating to bring good social results from the pursuit of self-interest. Under a free market system, people who pursue their own interests benefit society by developing

wealth, by making goods and performing services, and by promoting the progress of science, technology and industry. I happen to accept the utilitarian justification for capitalism, but I also recognize that society may need to impose various restrictions on free markets in order to preserve important social values, such as employee safety, public health, and social justice.

4. It is often said that the Bible claims that money is the root of all evil, but this statement is not quite accurate. 1 Timothy 6:10 states: "For the love of money is the root of all evil: which while some have coveted after, they have erred from the faith, and pierced themselves through with many sorrows." This passage warns of the corrupting influence of money but does not treat money as evil per se. Matthew 21:12–13 tells the story of how Jesus threw the money changers out of the temple. In this story, Jesus is not condemning money or free trade but he is seeking to prevent its corrupting influence on the church. Later on in the book of Matthew, Jesus was asked whether it was wrong to pay tribute to Caesar (as opposed to God) by having an image of Caesar on a coin. He responded, "render unto Caesar the things which are Caesar's; and unto God the things that are God's" (Matthew 22: 21). One way of interpreting this phrase is that Jesus is saying the money is not evil, only that it must have its proper place in society. Money is one of those things that may be given to Caesar (i.e., the government). Other things, such as one's heart, mind and soul, should be given to God.

5. Incidentally, these are the three basic policy options the President's Council on Bioethics (2002) considered to address concerns about reproductive and research cloning. The council appealed to a variety of slippery slope arguments to voice its ethical concerns related to cloning. The council was concerned that research and reproductive cloning would lead to various abuses, including exploitation of women, commodification of embryos, eugenics, the destruction of the traditional family, confusion about identity and individuality, and loss of respect for the value of human life.

6. This is a reformulation of the classic formula articulated by Judge Learned Hand in *United States v Carroll Towing* (1947), which has had a great deal of influence in negligence law. Judges view it as an informal rule for balancing benefits and risks rather than as a mathematical equation (Diamond, Levine, and Madden 2000).

7. Some would argue that one may assign a probability to any hypothesis. Hence, there is no difference between a hypothesis that is plausible and one that is probable. When we lack enough evidence to obtain objective probabilities, we can assign subjective probabilities (or educated guesses) to various hypotheses, given the evidence we have at hand, and then update these probabilities using Bayes' theorem as we obtain new evidence. However, I do not endorse the use of subjective probabilities in science because these probabilities would merely reflect personal biases rather than any rational assessment of the evidence. Although Bayesian methods can achieve convergence on objective probabilities in the long run, in the short-term there is not sufficient time or evidence to eliminate potential biases related to the use of subjective probabilities. For further discussion see Earman (1992).

8. The reasonable person is not the same thing as a "rational agent," which is a fiction used in economics, decision theory, philosophy, political science, and evolutionary

biology. A rational agent is also someone who follows various normative standards. For example, in decision theory, a rational agent is someone who obeys the rules of logic, adheres to the axioms of probability theory, conforms to principles of consistency and coherence in assigning probabilities to hypotheses, and maximizes his or her expected utility. One would expect a reasonable person would usually act rationally, but it might sometimes be reasonable to violate norms that govern rationality. For instance, a rational agent would not place a series of bets under conditions that would allow his bookie to make a "Dutch book" against him. A Dutch book is a betting scheme in which an opposing bettor, the bookie, places a series of bets against someone that allows that bookie to always win the sum total of all of the bets (Skyrms 1985). For instance, suppose the two people are betting on a collection of college football games. It seems to me that a reasonable person might sometimes bet under Dutch book conditions, provided that the benefits of engaging in the bets outweighed the risks. For instance, suppose that this person wants to develop a business relationship with the bookie and doesn't care whether he loses the bets. For further discussion of the relationship between rationality and reasonableness, see Audi (2001).

9. As we have noted several times, the European Parliament and the EPC do not allow patents on human beings. The PTO has not taken a stand on this issue, but its Director of Biological Patents, John Doll (2000), told me in a personal communication that the PTO would not award a patent on a whole human being or human embryo. The PTO rejected a patent application for a human-chimpanzee chimera (or "humanzee"), which was filed by Stuart Newman and supported by Jeremy Rifkin (Marshall 1999c). Newman filed the application in order to force the courts to rule on whether human beings are patentable. He has appealed the PTO's decision.

10. I would also regard a patent on a whole human being as an immoral violation of human dignity. Concerning a possible patent on a humanzee genome, the morality or immorality of this patent would depend on the status of the humanzee. If humanzees are (or would be) persons, then patents on humanzee genomes would be immoral.

CHAPTER 7

1. The word "anticommons" is meant to be the antonym of the commons, which is public land that existed in England prior to the enclosure movement of the seventeenth and eighteenth centuries. Before the enclosure of the commons by fences, people were allowed to use the common land for farming. After the enclosure movement, this public land had disappeared.

CHAPTER 8

1. This is not an uncontroversial assumption, since philosophers, political scientists, and policy analysts do not all agree on questions about justice in health care. For further discussion, see Daniels (1984), Kamm (1996), Englehart (1996).

CHAPTER 9

1. For further discussion, see Rollin (1995), Shiva (1996), Reiss and Straughan (1996), Rifkin (1983, 1985, 1998), Pinstrup-Anderson and Schioler (2001), Nelson (2001), Nestle (2001), Barboza (2000), Butler and Reichhardt (1999), and National Academy of Sciences (2002).

2. One might argue that my proposed farmer's exemption would violate the TRIPS agreement. One the other hand, the agreement gives countries the option of compulsory licensing for national emergencies or crises. In a country that is facing famine, the government could use this provision to license farmers to reproduce GM crops from seeds.

CHAPTER 10

1. Developing a procedure for ranking values and interests is not easy task either, since we live in a pluralistic society. Even though we cannot agree on how to rank various values and interests, we can still adopt rules and methods for allowing different stakeholders to participate in political decisions and in the formation of public policy. These rules and methods embody various ideals we associate with democracy, such as openness, transparency, publicity, submission to the rule of law, due process, and fairness. See Gutmann and Thompson (1996) and Rawls (1993) for more on procedures for resolving the conflicts of values inherent in a pluralistic society.

2. An expected utility is the product of a value and a probability. If I have a 25 percent chance of winning $4 at Bingo, a 50 percent chance of winning $2, and a 25 percent chance of winning $0, my expected utility = ($4 x .25) + ($2 x .50) + ($0 x .25) = $2.00. See Resnik (1987).

References

Adams, L. et al. 1999. Method for determining the presence of mutated BRCA protein. U.S. Patent 5,965,377. Washington, DC: U.S. Patent and Trademark Office.

Adams, M. et al. 2000. The genome sequence of drosophila melanogaster. *Science* 287:2185–95.

Alfred Bell & Co. v Catalda Fine Arts. 191 F. 2d 99 (2nd Cir. 1951).

American Heritage Dictionary of the English Language. 4th ed. 2000. New York: Houghton Mifflin.

American Society of Human Genetics (ASHG). 1991. Position paper on patenting of expressed sequence tags. Washington, DC: American Society of Human Genetics.

Anderson, E. 1990. Is women's labor a commodity? *Philosophy and Public Affairs* 19, no. 1:71–92.

Andre, J. 1992. Blocked exchanges: A taxonomy. *Ethics* 103, no. 1:29–47.

Andrews, E. 1995. Religious leaders prepare to fight patents on genes. *New York Times*, 13 May, N1, L1.

Andrews, L., and D. Nelkin. 2001. *Body bazaar*. New York: Crown Publishers.

Annas, G. 2000. Rules for research on human genetic variation: Lessons from Iceland. *Journal of the American Medical Association* 342:1830–33.

Application of Nelson. 280 F. 2d 172, 175 (Cust. & Pat. App. 1960).

Audi, R. 2001. *The architecture of reason*. New York: Oxford University Press.

Ayala, F. 1982. *Population and evolutionary genetics*. Menlo Park, CA: Benjamin Cummings.

Baggot, B. 1998. Human gene therapy patents in the United States. *Human Gene Therapy* 9, no. 1:151–57.

Baier, A. 1984. For the sake of future generations. In *Earthbound*, edited by T. Reagan. New York: Random House.

Baier, K. 1958. *The moral point of view.* Ithaca, NY: Cornell University Press.

Baker v Selden. 101 U.S. 99 (1879).

Balter, M. 2000. France rebels against gene-patenting law. *Science* 288:2115.

———. 2001. Tranatlantic war over BRCA1 patent. *Science* 292:1818.

Barron, J., and C. Dienes. 1999. *Constitutional law.* St. Paul, MN: West Group.

Barboza, D. 2000. Suburban genetics: Scientists searching for a perfect lawn. *New York Times,* 8 July, A1.

Barinaga, M. 1999. No winners in patent shootout. *Science* 284:1752–53.

Barton, J. 2000. Reforming the patent system. *Science* 287:1933–34.

Barton, J., and P. Berger. 2001. Patenting agriculture. *Issues in Science and Technology Online* 17, no. 4 (Summer): http://www.nap.edu/issues/17.4/p_barton.htm. Accessed: 15 February 2002.

Beardsley, T. 1994. Big-time biology. *Scientific American* 271, no. 5:90–94.

Benn, S. 1967. Property. In *Encyclopedia of philosophy,* edited by P. Edwards. New York: Macmillan.

Berenson, A., and N. Wade. 2000. A call for sharing of research causes gene stocks to fall. *New York Times,* 15 March, A1.

Berg, P. 1975. Summary statement of the Asilomar conference on recombinant DNA molecules. *Proceedings of the National Academy of Sciences* 72:1981–84.

Berman, H., D. Goodsell, and P. Bourne. 2002. Protein structures: From famine to feast. *American Scientist* 90:350–59.

Bernat, J., C. Culver, and B. Gert. 1982. Defining death in theory and practice. *Hastings Center Report* 12, no. 1:5–8.

Biotechnology Industry Organization (BIO). 2000. *Primer: Genomic and genetic research, patent protection, and twenty-first century medicine.* Washington, DC: Biotechnology Industry Organization.

———. 2001a. *The economic contributions of the biotechnology industry to the U.S. economy.* Washington, DC: Biotechnology Industry Organization.

———. 2001b. *State government initiatives in biotechnology in 2001.* Washington, DC: Biotechnology Industry Organization.

Black's Law Dictionary. 7th ed. 1999. St. Paul, MN: West Group.

Blaug, M. 1980. *The methodology of economics.* Cambridge: Cambridge University Press.

Blumenthal, D., E. Campbell, N. Causino, and K. Louis. 1996. Participation of life-science faculty in relationships with industry. *New England Journal of Medicine* 335:1734–39.

Blumenthal, D., E. Campbell, M. Anderson, N. Causino, and K. Louis. 1997. Withholding research results in academic life science: Evidence from a national survey of faculty. *Journal of the American Medical Association* 277:1224–28.

Bodenheimer, T. 2000. Conflict of interest in clinical drug trials: A risk factor for scientific misconduct. Paper presented at the Office of Human Research Protections conference on conflicts of interest in research, Washington, DC, 15 August.

Broad, W., and N. Wade. 1993. *Betrayers of truth*. 2nd ed. New York: Simon and Schuster.

Brody, B. 1999. Protecting human dignity and the patenting of human genes. In *perspectives on gene patenting*, edited by A. Chapman. Washington, DC: American Association for the Advancement of Science.

Brunner, E. 1947. *The divine imperative*. Philadelphia: Westminster Press.

Burke, J. 1995. *The day the universe changed*. Boston: Little, Brown.

Butler, D., and T. Reichhardt. 1999. Long-term effects of GM crops serves up food for thought. *Nature* 398:651–56.

Cahil, L. 2001. Genetics, commodification, and social justice in the globalization era. *Kennedy Institute for Ethics Journal* 11, no. 3:221–38.

Calamari, J., and J. Perillo. 1998. *The law of contracts*. 4th ed. St. Paul, MN: West Group.

Campbell, E. 2002. Personal communication, 23 January.

Campbell, E., B. Clarridge, M. Gokhale, L. Birenbaum, S. Hilgartner, N. Holtzman, and D. Blumenthal. 2002. Data withholding in academic genetics: Evidence from a national survey. *Journal of the American Medical Association* 287:473–80.

Caplan, A., and J. Merz. 1996. Patenting gene sequences: Not in the best interests of science or society. *British Medical Journal* 312:926.

Caulfield, T., and R. Gold. 2000a. Genetic testing, ethical concerns, and the role of patent law. *Clinical Genetics* 57:370–75.

———. 2000b. Whistling in the wind. *Forum for Applied Research and Public Policy* 15, no. 1:75–80.

Celera Genomics. 2001. Celera web page. www.celera.com. Accessed: 1 July 2001.

Chakrabarty, A. 1981. Microorganisms having multiple compatible degradative energy-generating plasmids and preparation thereof. U.S. Patent 4,259,444. Washington, DC: U.S. Patent and Trademark Office.

Chan, A., K. Chong, C. Martinovich, C. Simerly, and G. Schatten. 2001. Transgenic monkeys produced by retroviral gene transfer into mature oocytes. *Science* 291:309–12.

Chapman, L., and E. Bloom. 2001. Clinical xenotransplantation. *Journal of the American Medical Assocation* 285:2304–6.

Chea, T. 2002. Litter of gene-altered pigs cloned. *Washington Post,* 3 January, A1.

Christian Century. 1995. Call for a moratorium on gene patenting. *Christian Century* 112, no. 20:633–34.

Churchill, L. 1999. The United States health care system under managed care: How the commodification of health care distorts ethics and threatens equity. *Health Care Analysis* 7:393–411.

Claverie, J. 2001. What if there are only 30,000 human genes? *Science* 291:1255–57.

Clayton, E., K. Steinberg, M. Khoury, E. Thomson, L. Andrews, M. Kahn, L. Kopelman, and J. Weiss. 1995. Informed consent for stored tissue samples. *Journal of the American Medical Association* 274:1786–92.

Cohen, S. 1999. Selling bits and pieces of humans to make babies. *Journal of Medicine and Philosophy* 24:288–306.

Cohen, S., and H. Boyer. 1980. Process for producing biologically functional molecular chimeras. U.S. Patent 4,237,224.

Cole-Turner, R. 1997. Genes, religion, and society: The developing views of the churches. *Science and Engineering Ethics* 3:273–88.

———. 1999. Personal communication. 15 July 1999.

Collins, F., and V. McKusick. 2001. Implications of the human genome project for medical science. *Journal of the American Medical Association* 285:540–44.

Copi, I. 1986. *Introduction to logic,* 7th ed. New York: Macmillan.

Council on Ethical and Judicial Affairs, American Medical Association (AMA). 1997. *Patenting the human genome.* Chicago: American Medical Association.

———. 1998. *Current opinions with annotations.* Chicago: American Medical Association.

Council for Responsible Genetics (CRG). 2000. The genetic bill of rights. Cambridge, MA: Council for Responsible Genetics.

Cranor, C. 2001. Learning from law to address uncertainty in the precautionary principle. *Science and Engineering Ethics* 7, no. 3:313–26.

Crespi, R. 2000. An analysis of moral issues affecting patenting inventions in the life sciences: A European perspective. *Science and Engineering Ethics* 6, no. 2:157–80.

Crigger, B. 1995. The vampire project. *Hastings Center Report* 25, no. 1:2.

Daniels, N. 1984. *Just health care.* Cambridge: Cambridge University Press.

Dastgheib-Vinarov, S. 2000. A higher nonobviousness standard for gene patents: Protecting biomedical research from the big chill. *Marquette Intellectual Property Law Review* 4:143–79.

Davidson, R. 1986. Source of funding and outcome of clinical trials. *Journal of General Internal Medicine* 1:155–58.

DeAngelis, C. 2000. Conflict of interest and the public trust. *Journal of the American Medical Association* 284:2237–38.

Dellar v Samuel Goldwyn, Inc. 104 F. 2d 661, 662 (C.A.2 1939).

Derwent Information. 2001a. Frequently asked questions. http://www.derwent.com/. Accessed: 20 April 20.

———. 2001b. Glossary of patent terms. http://www.derwent.com/. Accessed: 20 April.

Diamond, J., L. Levine, and M. Madden. 2000. *Understanding torts.* 2nd ed. New York: Lexis Publishing.

Diamond v Chakrabarty, 447 U.S. 303 (1980).

Doll, J. 1998. The patenting of DNA. *Science* 280:689–90.

———. 2000. Personal communication. 1 September 2001.

Dreyfuss, R. 1989. General overview of the intellectual property system. In *Owning scientific and technical information,* edited by V. Weil and J. Snapper. New Brunswick, NJ: Rutgers University Press.

———. 2000. Collaborative research: Conflicts on authorship, ownership, and accountability. *Vanderbilt Law Review* 53:1161–1232.

Dreyfuss, R., and R. Kwal. 1996. *Intellectual property.* Westbury, NY: Foundation Press.

Earman J. 1992. *Bayes or bust?* Cambridge: MIT Press.

Eastman Kodak Co. v Image Technical Services, Inc. 112 S.Ct. 2072 (U.S.Cal.1992).

Eisenberg, R. 1990. Patenting the human genome. *Emory Law Journal* 39:721–45.

———. 1995. Patenting organisms. *Encyclopedia of bioethics.* Revised ed. New York: Simon and Schuster.

———. 1997. Structure and function in gene patenting. *Nature Genetics* 15, no. 2:125–30.

Englehart, T. 1996. *The foundations of bioethics.* 2nd ed. New York: Oxford University Press.

Enriquez, J. 1998. Genomics and the world's economy. *Science* 281:925–26.

Enserink, M. 1999. GM crops in cross hairs. *Science* 286:1662–68.

———. 2000. Patent office may raise the bar on gene claims. *Science* 287:1196–97.

European Commission. 1998. Opinions of the group of advisors on the ethical implications of biotechnology of the European Commission. Brussels: European Commission.

———. 2000. *Communication from the commission on the precautionary principle.* Brussels: European Commission.

————. 2001. The role of the European Commission. http://europa.eu.int/comm/
role_en.htm#1. Accessed: 1 June 2001.

European Parliament and Council. 1998. Directive 98/44/EC on the legal protection
of biotechnological inventions. *Official Journal L* 213, 30 (July): 13–21.

European Patent Convention (EPC). 1998. Convention on the grant of European
patents. Munich: European Patent Convention.

Evans, J. 1999. The uneven playing field on the dialogue on patenting. In *Perspectives
on gene patenting,* edited by A. Chapman. Washington, DC: American Association for the Advancement of Science.

Eweson, S. 1976. Apparatus for making organic fertilizer. U.S. Patent 3,930,799.
Washington, DC: U.S. Patent and Trademark Office.

Ex Parte Hibberd et al. 227 United States Patent Quarterly 443 (1985).

Feinberg, J. 1973. *Social philosophy.* Englewood Cliffs, NJ: Prentice-Hall.

Feinman, J. 2000. *Law 101.* New York: Oxford University Press.

Feist Publications Inc. v Rural Telephone Service Co., Inc. 499 U.S. 340 (1991).

Fields, S. 2001. Proteomics in genomeland. *Science* 291:1221–24.

Finkel, E., D. Malakoff, M. Balter, and A. Lawler. 2002. Patenting fight, round 2. *Science* 295:1621.

Fisher, L. 1999. The race to cash in on the genetic code. *New York Times,* 29 August,
A1.

Florida Prepaid Postsecondary Ed. Expenses Bd. v College Savings Bank. 527 U.S. 627
(1999).

Fons, M. 2000. The intellectual property landscape in the field of plasmid-based gene
therapy. *Journal of Drug Targets* 7:407–11.

Foster, F., and R. Shook. 1993. *Patents, copyrights, and trademarks.* 2nd ed. New York:
John Wiley.

Foster, F., P. Vecchia, and M. Repacholi. 2000. Science and the precautionary principle. *Science* 288:979–81.

Fox, D., and T. Paul. 1999. The "EST" dilemma: Should the discovery of a part of a
gene lead to a patent that could cover the entire gene once discovered? *Health
Law News* 13, no. 2:7.

Fox, R., and J. DeMarco. 1990. *Moral reasoning.* Chicago: Holt, Rinehart and Win-
ston.

Foubister V. 2000. Gene patents raise concerns for researchers, clinicians. *American
Medical News* (21 February): 1–4.

Frankena, W. 1973. *Ethics.* 2nd ed. Englewood Cliffs, NJ: Prentice-Hall.

Funk Brothers Seed Co. v Kalo Inoculant Co. 333 U.S. 127 (1948).

Gabriel, S. et al. 2002. The structure of haplotype blocks in the human genome. *Science* 296:2225–29.

Ganeri, A. 1997. *Religions explained.* New York: Henry Holt.

Gardner, W., and J. Rosenbaum. 1998. Database protection and access to information. *Science* 281:786–87.

Garfinkle, A. 1981. *Forms of explanation.* New Haven, CT: Yale University Press.

Garris, J. 1996. The case for patenting medical procedures. *American Journal of Law and Medicine* 22, no. 1:85–108.

Gellhorn, E., and W. Kovacic. 1994. *Antitrust law and economics.* 4th ed. St. Paul, MN: West Group.

Gelsinger Complaint. 2002. http://www.sskrplaw.com/links/healthcare2.html. Accessed: 15 August 2002.

Genentech Inc. v Chiron Corp. 112 F. 3d 495 (U.S. App. Ct., Fed. Cir. 1997).

Genetic Alliance. 2000. Statement submitted on gene patenting. Subcommittee on Courts and Intellectual Property, U.S. House of Representatives hearing on gene patents and other genomic inventions, 13 July 13.

Gibbard, A. 1990. *Wise choices, apt feelings.* Cambridge: Harvard University Press.

———. 1998. Which of our genes make us human? *Science* 281:1432–34.

Gibbs, W. 1996. The price of silence: Does profit-minded secrecy retard scientific progress? *Scientific American* 275, no. 5:15–16.

———. 1997. Plantibodies: How human antibodies produced by field crops enter clinical trials. *Scientific American* 277, no. 5:44.

Gill, M., and R. Sade. 2002. Paying for kidneys: The case against prohibition. *Kennedy Institute of Ethics Journal* 12, no. 1:17–45.

Gitter, D. 2001. International conflicts over patenting human DNA sequences in the United States and the European Union: An argument for compulsory licensing and a fair-use exemption. *New York University Law Review* 76:1623–89.

Glendon, M. 1991. *Rights talk.* New York: Free Press.

Gold, E. 1996. *Body parts: Property rights and the ownership of human biological materials.* Washington, DC: Georgetown University Press.

Gold, E., and T. Caulefield. 2002. The moral tollbooth: A method that makes use of the patent system to address ethical concerns in biotechnology. *Lancet* 359:2268–70.

Goldhammer, A. 2001. Current issues in clinical research and the development of pharmaceuticals. *Accountability in Research* 8:283–92.

Goozner, M. 2000. The price isn't right. *American Prospect* 11, no. 20:1.

Gorovitz, S. et al. 1979. *Philosophical analysis*. 3rd ed. New York: Random House.

Gosselin, P., and P. Jacobs. 2000. Patent office now at heart of debate. *Los Angeles Times*, 7 February, A1.

Graham v John Deere Co. 383 U.S. 1, 17–18 (1966).

Green, R. 2001. What does it mean to use someone as a means only: Rereading Kant. *Kennedy Institute of Ethics Journal* 11, no. 3:247–61.

Greenberg v Miami Children's Hosp. Research Institute, Inc. 2002 WL 1483266. (N.D.Ill. 2002).

Guenin, L. 1996. Norms for patents concerning human and other life forms. *Theoretical Medicine* 17:279–314.

Gura, T. 2001. After a setback, gene therapy progresses . . . gingerly. *Science* 291:1692–97.

Gutmann, A., and D. Thompson. 1996. *Democracy and disagreement*. Cambridge, MA: Harvard University Press.

Hamlyn, D. 1984. *Metaphysics*. Cambridge: Cambridge University Press.

Hanson, M. 1997. Religious voices in biotechnology: The case of gene patenting. *Hastings Center Report* 27, no. 6 (Special Supplement): 1–21.

———. 1999. Biotechnology and commodification within health care. *Journal of Medicine and Philosophy* 24, no. 3:267–87.

Hegel, F. 1976. [1821]. *The philosophy of the right*. Knox, T (trans.). Oxford: Oxford University Press.

Heller, M., and R. Eisenberg. 1998. Can patents deter innovation? The anticommons in biomedical research. *Science* 280:698–701.

Henry, M., M. Cho, M. Weaver, and J. Merz. 2002. DNA patenting and licensing. *Science* 297:1279.

Hess, D. 1999. Suppression, bias, and selection in science: The case of cancer research. *Accountability in Research* 6, no. 4:245–57.

Hettinger, E. 1989. Justifying intellectual property. *Philosophy and Public Affairs* 18:31–52.

Hill, T. 1992. *Dignity and practical reason in kant's moral theory*. Ithaca, NY: Cornell University Press.

Hoedemaekers, R., and W. Dekkers. 2002. Is there a unique moral status of human DNA that prevents patenting? *Kennedy Institute of Ethics Journal* 11, no. 4:359–86.

Holy Bible (King James Version). 1997. Nashville, TN: Thomas Nelson.

Honore, A. 1977. Ownership. In *The nature and process of law*, edited by P. Smith. New York: Oxford University Press.

Hughes, S. 2001. Making dollars out of DNA: The first major patent in biotechnology and the commercialization of molecular biology, 1974–1980. *Isis* 92:541–75.

Human Genome Organization (HUGO). 1995. Statement on the patenting of DNA sequences. Bethesda, MD: Human Genome Organization.

———. 2000. Statement on benefit sharing. Bethesda, MD: Human Genome Organization.

Human Genome Project. 2001. The science behind the human genome project. www.ornl.gov/hgmis/project/info.html. Accessed: 7 May 2001.

In re Deuel. 34 U.S. Pat. Quart. 1210 (2d Fed. Circ. 1995).

In re T.A.C.P. 609 S. 2d 588 (Florida 1992).

In re Wiggins. 488 F. 2d 538 (Cust. and Pat. App. 1973).

Israel, J., and W. LaFave. 1993. *Criminal procedure*. 5th ed. St. Paul, MN: West Group.

Jacob, M. 1997. *Scientific culture and the making of the industrial west*. New York: Oxford University Press.

Jacobson, M. et al. 1991. Surrogate markers for survival in patients with AIDS and AIDS related complex treated with zidovudine. *British Medical Journal* 302:73–78.

Joint Appeal against Human and Animal Patenting. 1995. Press Conference Text. Washington, DC: Board of Church and Society of the United Methodist Church, 17 May.

Juengst, E. 1998. Should we treat the human germ-line as a global human resource? In *Germ-line intervention and our responsibilities to future generations*, edited by E. Agius E. and S. Busuttil. London: Kluwer Academic Press.

Juengst, E., and M. Fossel. 2000. The ethics of ES cells—Now and forever, cells without end. *Journal of the American Medical Association* 284:3180–84.

Kahane, H. 1990. *Logic and philosophy: A modern introduction*. 6th ed. Belmont, CA: Wadsworth.

Kahn, P. 1998. Coming to terms with genes and risk. *Science* 274:496–500.

Kamm, F. 1996. *Morality, mortality*. Vol. 1 & 2. New York: Oxford University Press.

Kant, I. [1785] 1981. *Groundwork of the metaphysic of morals*. Translated by J. Ellington. Reprint, Indianapolis, IN: Hackett Publishing.

Kantorovich, A. 1993. *Scientific discovery*. Albany, NY: State University of New York Press.

Karp, J. 1991. Experimental use as patent infringement: The impropriety of broad exemption. *Yale Law Journal* 100:2169–88.

220 References

Kass, L. 1985. *Toward a more natural science: Biology and human affairs*. New York: Free Press.

Kaufman, S. 1993. *The origins of order*. New York: Oxford University Press.

Kevles, D. 2001. Patenting life: A historical overview of law, interests, and ethics. Paper prepared for Yale Legal Theory Workshop. Available at: http://www.yale.edu/law/ltw/papers/ltw-kevles.pdf. Accessed: 22 August 2002.

Kevles, D., and A. Berkowitz. 2001. The gene patenting controversy: A convergence of law, economic interests, and ethics. *Brooklyn Law Review* 67:233–48.

Kimbrell, A. 1997. *The human body shop*. Washington, DC: Gateway.

Kitcher, P. 1993. *The advancement of science*. New York: Oxford University Press.

———. 1997. *The lives to come*. New York: Simon and Schuster.

Kiwanee Oil Co. v Bicron Corp. 416 U.S. 470, 480 (1974).

Kleyn, P., and E. Vesell. 1998. Genetic variation as a guide to drug development. *Science* 281:1820–21.

Knoppers, M. 1999. Status, sale, and patenting of human genetic material: An international survey. *Nature Genetics* 22:23–26.

Knoppers, B., M. Hirtle, and K. Glass. 1999. Commercialization of genetic research and public policy. *Science* 286:2277–78.

Ko, Y. 1992. An economic analysis of biotechnology patent protection. *Yale Law Journal* 102:777–802.

Kolata, G. 2000. Who owns your genes? *New York Times*, 15 May, A1.

Korn, D. 2000. Conflict of interest in biomedical research. *Journal of the American Medical Association* 284:2234–37.

Korsgaard, C. 1996. *The sources of normativity*. Cambridge: Cambridge University Press.

Kosso, P. 1992. *Reading the book of nature*. Cambridge: Cambridge University Press.

Krimsky, S., L. Rothenberg, P. Stott, and G. Kyle. 1996. Financial interests of authors in scientific publications. *Science and Engineering Ethics* 2, no. 4:395–410.

Kuflik, A. 1989. Moral foundations of intellectual property rights. In *Owning scientific and technical information,* edited by V. Weil and J. Snapper. New Brunswick, NJ: Rutgers University Press.

Kuhn, T. 1962. *The structure of scientific revolutions*. Chicago: University of Chicago Press.

Lafollette, N., and N. Shanks. 1996. *Brute science: Dilemmas of animal experimentation*. New York: Routledge.

Land, R., and B. Mitchell. 1996. Patenting life: No. *First Things* 63 (May): 20–22.

Langreth, R., and B. Davis. 2000. Press briefing set off rout in biotech. *Wall Street Journal*, 16 March, A19.

Lao, M. 1999. Unilateral refusals to sell or license intellectual property and the antitrust duty to deal. *Cornell Journal of Law and Public Policy* 9:193–219.

Laudan, L. 1978. *Progress and its problems*. Berkeley: University of California Press.

Leonard, D. 1999. The future of molecular genetic testing. *Clinical Chemistry* 45:726–31.

Locke, J. [1764] 1980. *Second treatise of government*. Reprint, Indianapolis, IN: Hackett.

Looney, B. 1994. Should genes be patented? The gene patenting controversy: Legal, ethical, and policy foundations of an international agreement. *Law and Policy in International Business* 26:231–72.

Lucotte, G., and F. Baneyx. 1993. *Introduction to molecular cloning techniques*. New York: VCH Publishers.

Lyons, D., ed. 1979. *Rights*. Belmont, CA: Wadsworth.

Mackie, J. 1977. *Ethics: Inventing right and wrong*. London: Penguin Books.

Malakoff, D. 2000. The rise of the mouse, biomedicine's model animal. *Science* 288:48–53.

Malakoff, D., and E. Marshall. 1999. NIH wins big as Congress lumps together eight bills. *Science* 282:598.

Marshall, E. 1997. Snipping away at genome patenting. *Science* 277:1752–53.

———. 1999a. Drug firms to create public database of genetic mutations. *Science* 284:406–7.

———. 1999b. A high-stakes gamble on genome sequencing. *Science* 284:1906–9.

———. 1999c. Legal fights over patents on life. *Science* 294:2067–68.

———. 2000a. Patent on HIV receptor provokes outcry. *Science* 287:1375–77.

———. 2000b. Talks of public-private deal end in acrimony. *Science* 287:1723–24.

———. 2000c. A deluge of patents creates legal hassles for research. *Science* 288:255–57.

———. 2000d. Claim and counterclaim on the human genome. *Science* 288:242–43.

———. 2000e. In the crossfire: Collins on genomes, patents, and rivalry. *Science* 287:2396–98.

———. 2000f. Gene therapy on trial. *Science* 288:951–57.

———. 2001. Sharing the glory, not the credit. *Science* 291:1189–93.

Marx, K. [1867] 1996. *Das Kapital*. Reprint, New York: Regnery Publishing.

Marx, K., and F. Engels. [1848] 1998. *The communist manifesto*. Reprint, New York: Signet Books.

Maughan, J., F. Lewis, and V. Smith. 2001. An introduction to arrays. *Journal of Pathology* 195, no. 1:3–6.

Mayr, E. 1982. *The growth of biological thought*. Cambridge: Harvard University Press.

Mazer v Stein. 347 U.S. 201 (1954).

McCrary, V. et al. 2000. A national survey of policies of disclosure of conflict of interest in biomedical research. *New England Journal of Medicine* 343:1621–26.

McDowell, N. 2002. Africa hungry for traditional food as biotech row drags on. *Nature* 418:571–72.

McQueen, M. 1998. Ethical and legal issues in the procurement, storage and use of DNA. *Clinical chemistry and laboratory medicine* 36:545–49.

Meadows, J. 1992. *The great scientists*. New York: Oxford University Press.

Mehlman, M., and J. Botkin. 1998. *Access to the genome*. Washington, DC: Georgetown University Press.

Merck & Co., Inc. v Olin Mathieson Chemical Corporation. 253 F. 2d 156 (4th Circ. 1958).

Merz, J. 2000. Statement submitted on gene patenting. Subcommittee on courts and intellectual property. U.S. House of Representatives hearing on gene patents and other genomic inventions, 13 July.

Merz, J., and M. Cho. 1998. Disease genes are not patentable: A rebuttal to McGee. *Cambridge Quarterly of Healthcare Ethics* 7:425–28.

Merz, J., M. Cho, M. Robertson, and D. Leonard. 1997. Disease gene patenting is bad innovation. *Molecular Diagnosis* 2, no. 4:299–304.

Merz, J., D. Magnus, M. Cho, and A. Caplan. 2002. Protecting subjects' interests in genetics research. *American Journal of Human Genetics* 71, no. 4:965–71.

Mill, J. [1859] 1956. *On liberty*. Reprint, New York: Liberal Arts Press.

——. [1861] 1979. *Utilitarianism*. Reprint, Indianapolis, IN: Hackett.

Miller, A., and M. Davis. 2000. *Intellectual property: Patents, trademarks, and copyright*. St. Paul, MN: West Group.

Mitchell, C. 1999. A Southern Baptist looks at patenting life. In *Perspectives on gene patenting*, edited by A. Chapman. Washington, DC: American Association for the Advancement of Science.

Moore v Regents of the University of California. 793 P.2d 479 (Cal. 1990).

Morelli, M. 1999. Commerce in organs: A Kantian critique. *Journal of Social Philosophy* 30:315–24.

Morin, K. et al. 2002. Managing conflicts of interest in the conduct of clinical trials. *Journal of the American Medical Association* 287:78–84.

Morreim, H. 1995. *Balancing act: The new medical ethics of medicine's new economics.* Washington, DC: Georgetown University Press.

Munthe, C., and S. Welin. 1996. The morality of scientific openness. *Science and Engineering Ethics* 2, no. 4:411–28.

Murray, T. 1986. Who owns the body? On the ethics of using human tissue for commercial purposes. *IRB: A Review of Human Subjects Research* 8, no. 1:1–5.

Nader, R. 2000. Nader 2000 http://www.votenader.org/issues/agriculture_policy.html. Accessed: 6 November.

Nagel, E. 1977. Teleology revisited. *Journal of Philosophy* 74:261–302.

National Academy of Sciences (NAS). 1996. *Guide for the care and use of laboratory animals.* Washington, DC: National Academy of Sciences.

———. 2002. *Animal biotechnology: Identifying science-based concerns.* Washington, DC: National Academy Press.

National Commission for the Protection of Human Subjects in Biomedical and Behavioral Research. 1979. *The Belmont report.* Washington, DC: National Commission for the Protection of Human Subjects in Biomedical and Behavioral Research.

National Bioethics Advisory Commission (NBAC). 1998. *Research involving human biological materials: Ethical issues and policy guidance.* Washington, DC: National Bioethics Advisory Commission.

National Institutes of Health (NIH). 2001. Press Release for the FY 2002 President's Budget." Bethesda, MD: National Institutes of Health.

———. 2002. Press Release for FY 2003 President's Budget. Bethesda, MD: National Institutes of Health.

Nelkin, D., and L. Andrews. 1998. Homo economicus: The commercialization of body tissue in the age of biotechnology. *Hastings Center Report* 28, no. 5:30–39.

Nelkin, D., and S. Lindee. 1995. *The DNA mystique.* New York: Free Press.

Nelson, G., ed. 2001. *Genetically modified organisms in agriculture: Economics and politics.* New York: Academic Press.

Nestle, M. 2001. Food biotechnology: Whose values, whose decisions? *Witness* (May): 15–17.

Newton-Smith, W. 1981. *The rationality of science.* London: Routledge.

Nickel, L. 1976. Ripening of sugarcane by use of certain alcoholic and ethoxylated compounds. U.S. Patent 3,930,840. Washington, DC: U.S. Patent and Trademark Office.

Niiniluoto, I. 1999. *Critical scientific realism*. New York: Oxford University Press.

Normile, D. 2000. Monsanto donates its share of golden rice. *Science* 289:843–44.

Nozick, R. 1974. *Anarchy, state, utopia*. New York: Basic Books.

Nuffield Council on Bioethics. 2002. *The ethics of patenting DNA*. London: Nuffield Council.

Olson, R. 1967. Deontological ethics. In *Encyclopedia of Philosophy*. Vol. 1 & 2. Edited by P. Edwards. New York: Macmillan.

O'Neill, O. 1996. *Towards justice and virtue: A constructive account of practical reason*. Cambridge: Cambridge University Press.

Ossario, P. 1999. Common heritage arguments and the patenting of DNA. In *Perspectives on gene patenting*, edited by A. Chapman. Washington, DC: American Association for the Advancement of Science.

Paabo, S. 2001. The human genome and our view of ourselves. *Science* 291:1219–20.

Pabst, P. 1999. Gene therapy and tissue engineering patents abound. *Tissue Engineering* 5, no. 1:79.

Parens, E. 1996. Taking behavioral genetics seriously. *Hastings Center Report* 26, no. 4:13–18.

Parker v Flook. 437 U.S. 584 (1978).

Patent and Trademark Office (PTO). 1999. Revised utility examination guidelines. *Federal Register* (21 December) 64, no. 244:71440–42.

———. 2001. General information regarding patents. Washington, DC: Patent and Trademark Office.

———. 2002. Patent search. www.uspto.gov. Accessed: 17 January 2002.

Pearl, J. 2000. *Causality*. Cambridge: Cambridge University Press.

Peltonen, L., and V. McKusick. 2001. Dissecting human disease in the postgenomic era. *Science* 291:1224–29.

Pennisi, E. 2000a. Stealth genome rocks researchers. *Science* 288:239–40.

———. 2000b. Fruit fly genome yields data and validation. *Science* 287:1374.

Peters, T. 1997. *Playing God*. New York: Routledge.

Pinstrup-Andersen, P., and E. Schioler. 2001. *Seeds of contention: World hunger and the global controversy over GM (genetically modified) crops*. Baltimore, MD: Johns Hopkins University Press.

Plant Variety Protection Act. 1994. 7 U.S.C. 2401.

Pojman, L. 1995. *Ethics*. 2nd ed. Belmont, CA: Wadsworth.

Poland, S. 2000. Genes, patents, and bioethics: Will history repeat itself? *Kennedy Institute of Ethics Journal* 10, no. 3:265–81.

Pollack, A. 2002. Genome pioneer will start center of his own. *New York Times,* 16 August, C1.

Popper, K. 1959. *The logic of scientific discovery.* London: Hutchinson.

Porter, R. 1997. *The greatest benefit of mankind.* New York: W.W. Norton.

President's Commission for the Study of Ethical Problems in Medicine and Biomedical and Behavioral Research. 1981. *Defining death.* Washington, DC: President's Commission.

President's Council on Bioethics. 2002. *Human cloning and human dignity: An ethical inquiry.* Washington, DC: President's Council on Bioethics.

Price, S., and L. Wilson. 1982. *Pathophysiology.* 2nd ed. New York: McGraw-Hill.

Public Health Service (PHS). 2000. Draft interim guidance: Financial relationships in clinical research. Washington, DC: Public Health Service.

Rachels, J. 1993. *The elements of moral philosophy.* 2nd ed. New York: McGraw-Hill.

Radin, M. 1996. *Contested commodities.* Cambridge: Harvard University Press.

Rawls, J. 1971. *A theory of justice.* Cambridge: Harvard University Press.

———. 1993. *Political liberalism.* New York: Columbia University Press.

Regents of the University of California v Eli Lilly and Co. 119 F. 3d 1159 (U.S. Ct. App. Fed. Cir. 1997)

Reichhardt, T. 1998. Patent on gene fragment sends researchers a mixed message. *Nature* 396:499.

Reiss, M., and R. Straughan. 1996. *Improving nature? The science and ethics of genetic engineering.* Cambridge: Cambridge University Press.

Resnik, D. 1993. Do scientific aims justify methodological rules? *Erkenntnis* 38:223–32.

———. 1996. Basic research and corporate responsibility. *Business and Society Review* 96:57–59.

———. 1997. The morality of human gene patents. *Kennedy Institute of Ethics Journal* 7, no. 1:43–61.

———. 1998a. The commodification of human reproductive materials. *Journal of Medical Ethics* 24:388–93.

———. 1998b. *The ethics of science.* New York: Routledge.

———. 1998c. Industry sponsored research: Secrecy versus corporate responsibility. *Business and Society Review* 99:31–34.

———. 1999a. Privatized biomedical research, public fears, and the hazards of government regulation: Lessons from stem cell research. *Health Care Analysis* 7:273–87.

———. 1999b. The human genome diversity project: Ethical problems and solutions. *Politics and the Life Sciences* 18, no. 1:15–23.

———. 1999c. Scientific rationality and epistemic goals. *Protosoziologie* 12:258–89.

———. 2000. Financial interests and research bias. *Perspectives on Science* 8, no. 3:255–85.

———. 2001a. DNA patents and scientific discovery and innovation: Assessing benefits and risks. *Science and Engineering Ethics* 7, no. 1:29–62.

———. 2001b. DNA patents and human dignity. *Journal of Law, Medicine, and Ethics* 29, 2: 152.

———. 2001c. Developing drugs for the developing world: An economic, legal, moral, and political dilemma. *Developing World Bioethics* 1, no. 1:11–32.

———. 2001d. Regulating the market for human eggs. *Bioethics* 15, no. 1:1–25.

———. 2002a. Discoveries, inventions, and gene patents. In *Who owns the life?* edited by D. Magnus. New York: Prometheus Press.

———. 2003a. Is the precautionary principle unscientific? *Studies in the History and Philosophy of the Biological Sciences* 34:327–44.

———. 2003b. Exploitation in biomedical research. *Theoretical Medicine and Bioethics* 24:233–59.

Resnik, D., and K. De Ville. 2002. Bioterroism and patent rights: Compulsory licensure and the case of Cipro. *American Journal of Bioethics* 2, no. 3:29–39.

Resnik, D., and A. Shamoo. 2002. Conflict of interest and the university. *Accountability in Research* 9, no. 1:45–64.

Resnik, D., H. Steinkraus, and P. Langer. 1999. *Human germline gene therapy*. Austin, TX: RG Landes.

Resnik, M. 1987. *Choices: An introduction to decision theory*. Minneapolis: University of Minnesota Press.

Reynolds, T. 2000. Gene patent race speeds ahead amid controversy, concern. *Journal of the National Cancer Institute*, 92, no. 3:184–86.

Rifkin, J. 1983. *Algeny*. New York: Penguin Books.

———. 1985. *Declaration of a heretic*. London: Routledge and Kegan Paul.

———. 1998. *The biotech century*. New York: Penguin Putnam.

———. 2000. Personal communication, 17 July 2000.

Risjord, M. 2000. *Woodcutters and witchcraft*. Albany, NY: State University of New York Press.

Roberts, L. 2000. SNP mappers confront reality and find it daunting. *Science* 287:1898–99.

———. 2001. Controversial from the start. *Science* 291:1182–88.

Roberts, R. et al. 2001. Building a "Genbank" of the published literature. *Science* 291:2318–19.

Robertson, M., ed. 1994. *Molecular biology of the cell*. New York: Garland Publishing.

Roch Products v Bolar Pharmaceutical Co. 733 F. 2d 858 (Fed. Cir. 1984).

Rodwin, M. 1993. *Medicine, money and morals*. New York: Oxford University Press.

Rohwer, C., and A. Skrocki. 2000. *Contracts*. 5th ed. St. Paul: West Group.

Rollin, B. 1995. *The Frankenstein syndrome*. Cambridge: Cambridge University Press.

Rolston, H. 1994. *Conserving natural value*. New York: Columbia University Press.

Rosenberg, A. 1985. *The philosophy of biology*. Cambridge: Cambridge University Press.

———. 1995. *The philosophy of social science*. 2nd ed. Boulder, CO: Westview Press.

Ross, W. 1930. *The right and the good*. Oxford: Clarendon Press.

Rothstein, M., ed. 1997. *Genetic secrets*. New Haven, CT: Yale University Press.

Sackett, D., S. Straus, W. Richardson, W. Rosenberg, and B. Hayes. 1997. *Evidence-based medicine*. New York: Churchill Livingstone.

Sagoff, M. 1999. DNA patents: Making ends meet. In *Perspectives on gene patenting*, edited by A. Chapman. Washington, DC: American Association for the Advancement of Science.

Samuelson, P. 1980. *Economics*. 11th ed. New York: McGraw-Hill.

Saunders, J., and C O'Malley, eds. and trans. 1982. *The anatomical drawings of Andreas Vesalius*. New York: Bonanza Books.

Schatz, G. 1998. The Swiss to vote on gene technology. *Science* 281:1810–11.

Scheffler, S. 1988. *Consequentialism and its critics*. Oxford: Oxford University Press.

Schissel, A., J. Merz, and M. Cho. 1999. Survey confirms fears about licensing of genetic tests. *Nature* 402:118.

Schonmann, A. 1998. The patenting challenges of gene discovery. *Nature Biotechnology* 16 (Supplement): 47.

Scott, R. 2000. Testimony before the House Judiciary Subcommittee on courts and intellectual property, 13 July.

Service, R. 2000. Can Celera do it again? *Science* 287:2138–39.

———. 2001. Gene and proteins patents get ready to go head to head. *Science* 294:2082–83.

Shand, H. 1999. Terminator technology: Genetically altered seeds will destroy both diversity and good-producing capacity. *Christian Social Action* (October): 7–13.

Shattuck-Eidens, D. et al. 1997. Linked breast and ovarian cancer susceptibility gene. U.S. Patent 5,693,473. Washington, DC: U.S. Patent and Trademark Office.

Sharp, R., and M. Foster. 2001. Involving study populations in the review of research. *Journal of Law, Medicine and Ethics* 28, no. 1:41–49.

Shiva, V. 1996. *Biopiracy: The plunder of nature and knowledge.* Boston: South End Press.

Shoemaker, R. et al. 2001. *Economic issues in agricultural biotechnology.* Washington, DC: U.S. Department of Agriculture.

Shrader-Frechette, K. 1991. *Risk and rationality.* Berkeley: University of California Press.

Skyrms, B. 1985. *Choice and chance.* Belmont, CA: Wadsworth.

Smith, A. [1776] 1991. *The wealth of nations.* New York: Prometheus Books.

Smith, W. 2001. *The culture of death.* New York: Encounter Books.

Skolnick, M. et al. 1998. 170-linked breast and ovarian cancer susceptibility gene. U.S. Patent 5,753,441. Washington, DC: U.S. Patent and Trademark Office.

Stolberg, S., and J. Gerth. 2000. Medicine merchants: Holding down the competition. *New York Times,* 23 July, A1.

Sturges, M. 1997. Who should hold property rights to the human genome? An application of the common heritage of mankind. *American University International Review* 13:219–61.

Subramanian, G., M. Adams, J. Venter, and S. Broder. 2001. Implications of the human genome for understanding human biology and medicine. *Journal of the American Medical Association* 286:2296–2307.

Suzuki, D., and P. Knudtson. 1989. *Genethics: The clash between the new genetics and human values.* Cambridge: Harvard University Press.

Svatos, M. 1996. Biotechnology and the utilitarian argument for patents. *Social Philosophy and Policy* 13:113–44.

Swisher, K., ed. 1995. *What is sexual harassment?* San Diego, CA: Greenhaven Press.

Taubes, G. 1995. Scientists attacked for patenting Pacific tribe. *Science* 270:1112.

Telecomm Technical Services, Inc. v Siemens Rolm Communications, Inc. 150 F.Supp.2d 1365 (N.D.Ga. 2000).

Time Magazine, March 9, 1981.

Tribble, J. 1998. Gene patents—A pharmaceutical perspective. *Cambridge Quarterly of Healthcare Ethics* 7:429–32.

United Kingdom Patent Office (UKPO). 2001. Five hundred years of patents. London: United Kingdom Patent Office.

United Nations Educational, Scientific and Cultural Organization (UNESCO). 1997. Universal Declaration on the Human Genome and Human Rights. Geneva: United Nations Educational, Scientific and Cultural Organization.

United States v Carroll Towing Co. 159 F. 2d 169 (2d Circuit 1947).

United States v White Motor Co. 194 F.Supp. 562 (D.C.Ohio 1961).

U.S. Constitution. Article 1, Section 8, Clause 8 (1787).

U.S. Department of Agriculture (USDA). 2001. *Economic issues in agricultural biotechnology.* Washington, DC: U.S. Department of Agriculture.

U.S. Patent Act. 1995. 35 U.S.C. 101.

Vacchiano, E. 1999. It's a wonderful genome: The written description requirement protects the human genome from overly broad patents. *John Marshall Law Review* 32:805–32.

Venter, C. et al. 2001. The sequence of the human genome. Science 291:1304–51.

Volti, R. 1995. *Society and technological change.* 3rd ed. New York: St. Martin's Press.

Wade, N. 2000a. Analysis of genome is said to be complete. *New York Times,* 7 April, A1.

———. 2000b. Rivals in race to decode human DNA agree to cooperate. *New York Times,* 22 June, A1.

———. 2000c. Genetic code of human life is cracked by scientist. *New York Times,* 27 June, A1.

Waldron, J. 1988. *The right to private property.* Oxford: Clarendon Press.

Walter, J. 1982. *Principles of disease.* Philadelphia: W.B. Saunders.

Walters, L., and J. Palmer. 1997. *The ethics of human gene therapy.* New York: Oxford University Press.

Walzer, M. 1983. *Spheres of justice.* New York: Basic Books.

Wambugu, F. 1999. Why Africa needs agricultural biotech. *Nature* 400:15–16.

Warner-Jenkinson Co. v Hilton Davis Chemical Co. 520 U.S. 17 (1997).

Wertheimer, R. 1996. *Exploitation.* Princeton, NJ: Princeton University Press.

Whittemore v Cutter. 29 F. Ca. 1120 (C.C.D. Mass. 1813).

Wicklegren, I. 1999. Mining the genome for drugs. *Science* 285:998–1001.

———. 2000. Mutation points way to salt recycling pathway. *Science* 289:23–26.

Wilmut, I. 1998. Cloning for medicine. *Scientific American* 279 no. 6:58–63.

Wong, D. 1997. *The ABCs of gene cloning.* New York: Chapman and Hall.

Wood, R., M. Mitchell, J. Sgouros, and T. Lindahl. 2001. Human DNA repair genes. *Science* 291:1284–89.

Woollett, G., and O. Hammond. 1999. An industry perspective on the gene patenting debate. In *Perspectives on gene patenting*, edited by A. Chapman. Washington, DC: American Association for the Advancement of Science.

World Medical Association. 2001. Declaration of Helsinki. *Journal of the American Medical Association* 284:3043–46.

Wynia, M., S. Latham, A. Kao, J. Berg, and L. Emmanuel. 1999. Medical professionalism in society. *New England Journal of Medicine* 341:1612–16.

Yaphe, J., R. Edman, B. Knishkowy, and J. Herman. 2001. The association between funding by commercial interests and study outcome in randomized controlled drug trials. *Family Practice* 18, no. 6:565–68.

Zurer, P. 1994. NIH drops bid to patent gene fragments. *Chemical and Engineering News* 72, no. 8:5–6.

Index

accessibility, 170–74, 185–88
agrarian societies, 193
agriculture, 9,10, 72, 177–79, 193, 196, 197
American Medical Association (AMA), 156
Amgen, 54, 55, 68–71, 89, 174
Anderson, F., 160
Andrews, L., 94
animal welfare, 83, 182
anti-commons,146
antitrust laws, 43, 69, 152, 172, 173, 184

bacteria, 27, 29, 52
Bayh-Dole Act, 70
benefit sharing, 81, 165, 200
bioinformatics, 23
bioprospecting, 190, 191, 200
biotechnology industry, 6, 9, 10, 52, 67, 68, 80 134, 146, 186, 202
Boyer, H., 52
breast cancer, 111, 173
BRCA 1/BRCA II, 23, 25, 70, 159–60, 173

Canavan's disease, 158–59
Caplan, A., 161
categorical imperative, 95, 102
causation, 86–88, 205
Celerea Genomics, 2, 4–6, 45, 49, 51, 67, 68, 71, 150
Chakrabarty, A., 52–53
Christianity, 97

chromosomes, 14, 15,17
cloning, 27, 62, 119,179
Cohen, S., 52
Collins, F., 4
commercialization, 31, 32, 79, 91, 101, 105, 114, 168
common heritage argument, 10, 77–83
commodification, 10, 31, 32, 82–83, 93, 100, 103, 113–17, 120, 125–26, 170, 193
common property, 77–82
confidentiality, 141
conflict of interest, 144–46, 162, 166–68, 200
consequentialism, 8, 63, 79, 195
consumerism, 169
copyrights, 34, 46–51, 200
    creativity, 47
    fair use, 45, 48, 200
    infringement, 47–48
    on DNA, 51
    on facts or ideas, 47
    originality, 46, 49
    utility, 47
Council for Responsible Genetics, 3, 74, 77, 94
Crick, F., 14

Darwin, C. 139
databases, 49, 199
decision theory, 108
deontology, 9, 63, 75, 78, 88, 93, 195, 203

deoxyribonucleic acid (DNA), 14–28,
    78, 91, 114, 197, 199
  chips, 25
  cloning, 27
  complementary, 89
  human, 204–205
  junk, 17, 22, 59
  mutations, 15,16
  patents on (see patents, on DNA)
  recombinant, 52
  sequencing, 4–5, 22, 56
  transcription, 18
  translation, 18
developing nations, 80, 178, 186–88,
    190–91, 199
Diamond v. Chakrabarty, 29, 52–54, 84,
    85, 185
downstream effects, 43, 143, 146, 147,
    184
drug development, 24
dualism, 118
duties, 103

economics, 6, 8–9, 67–68
egalitarianism, 79–80
Eli Lilly, 190
eminent domain, 42
erythropoietin, 54, 70, 89
Ethical review, 200
European Commission (EC), 62, 121
European Patent Convention (EPC), 5,
    61, 89, 94
European Union (EU), 5, 62
Evolution, 18
Ex Parte Hibberd et al, 185
experimental use exemption (see
    research, exemption)
explanation, 87
exploitation, 1, 128, 162–66, 189–93
expressed sequence tags (ESTs), 3, 28

farmer's exemption, 185–88, 199
Feist Publications, Inc. vs. Rural Telephone
    Service, Inc., 49
Food and Drug Administration (FDA),
    109–10, 161
freedom, 36, 66, 96

free market, 149–50, 171–72, 206
function, 58–59
functional equivalence, 43, 59
Funk Brothers Seed Co. v. Kalo Inoculant
    Co, 52

Gelsinger, J., 160–62, 167
Genbank, 150
gene(s), 18–26 (see also deoxyribonucleic
    acid)
  Expression, 18
  Therapy, 26, 160–62
Genentech, 2, 44, 52, 68
generic drugs, 43, 44, 171, 172
genetic
  determinism, 22, 90
  diseases, 16, 20–23
  diversity, 17, 82, 182
  identity , 7, 117–22
  predispositions, 23
  testing, 25, 174–75
Genetic Alliance, 156
genetically modified
  animals, 7, 30, 179–82
  crops, 7, 30, 178–81, 185–88, 193
  foods, 30, 178–80, 187
genetics, 6, 13, 25, 136, 139
genome, 15
  human, 4, 5, 17, 78, 93, 95, 128, 198,
    204–205
genomics, 6, 22, 136, 139
genotypes, 13
Genovo, 162
God, 10, 75–77, 180
  image of, 97
Green, R., 102–103

Hagahai, 1–2, 157–58
harm principle, 35, 37–38
Harvard University, 29
Hegel, F., 36, 204
hemoglobin, 20
hemophilia, 21
homology, 28, 114
human dignity, 10, 66, 94, 99, 100, 105,
    106, 112–22, 163, 195
human genome (see genome, human)

Human Genome Diversity Project
  (HGDP), 1
Human Genome Organization
  (HUGO), 61, 77
Human Genome Project (HGP), 4, 22,
  71
human germ-line modification, 121,
  183
humanzee, 128, 201

ideas, 85
Incyte Pharmaceuticals, 2, 67, 150
indigenous people, 1, 3
industrialization, 193
informed consent, 157, 163, 165, 190,
  200
intellectual property, 32–39, 132, 140,
  191–92
  vs. tangible property, 32
  treaties, 60–61
invention, 39, 62, 87–88
  copycat, 149–50
  pioneer, 44
inventor, 39
Islam, 97
isolation and purification, 28, 54, 62,
  84–85, 113

Jefferson, T., 34, 53
Jesus, 98–99
Joint Appeal Against Human and
  Animal Patenting, 2, 74, 83, 93
Judaism, 97
Justice, 79–82, 163–65, 170–74, 185–89,
  190, 200

Kant, I., 91, 95–96, 99, 102, 206
Kimbrell, A., 93
Kuhn, T., 132

legal issues, 8, 53
licenses, 41, 147–52, 184
  compulsory, 42, 65, 148, 174
  cross-licenses, 42
  exclusive, 41
  non-exclusive, 41
  reach through licenses, 42

life
  patents on (see patents, on life forms)
  value of, 91
love, 98

market values, 32, 93, 100–101, 116
Marx, K. 206
maxim rule, 109
Merck and Company, 190
Merck & Co. v. Olin Mathieson Chemical
  Corp., 85
Medicine, 6, 9, 10, 23, 73, 155–56,
  196–97
Medline, 136–37
Mendel, G., 13
Miami Children's Hospital, 158
Mill, J., 8
molecular biology, 14, 136, 139
monopolies, 69, 151, 171–72, 184
Monsanto, 2, 149–50, 185
Moore, J., 138, 156–57, 163
Moore v. Regents of the University of
  California, 128, 156–57, 165
moral
  agency, 96
  analysis, 8, 63
  argument, 63
  rights, 65, 203
  rules, 96
  worth, 96, 101
morality, 62, 94
Myriad Genetics, 2, 54, 69–71, 159–60,
  170–71, 173

Nader, R., 94
National Human Genome Research
  Institute (NHGRI), 4
National Institutes of Health (NIH), 1,
  2, 70, 135, 159–60
natural vs. artificial, 85–89, 114
Nelkin, D., 94

objectivity, 86–88, 144, 196
oncomouse, 29
openness, 139–40, 197
ownership, 32, 113

part vs. whole, 115–17
patent(s), 33, 39–46, 140
    application, 39, 142–43, 199
    assignment, 41
    blocking, 149–50
    due diligence, 40
    enabling description, 40
    infringement, 42–43, 185
    novelty, 40, 140
    non-obviousness, 40, 56, 152, 198
    on bacteria, 29
    on computer programs, 84
    on DNA, 1–11, 51, 113, 136–38
    on genes, 62, 128
    on genetic tests, 174–75
    on human beings, 62, 93
    on improvements, 39
    on life forms, 62, 75, 93
    on processes, 39, 174–75
    on products, 39, 55
    scope, 44
    subject matter, 39–40
    utility, 4, 40, 57, 58, 152, 173, 198
patentability, 39–40, 53
Patent Act, 39, 53, 64
Patent and Trademark Office (PTO), 1,
    4, 34, 41, 43, 53, 54, 57, 89, 136, 138,
    161, 203
Perkin Elmer, 2, 22
Person, 95, 96, 99, 116–20
personal identity, 95, 96, 99, 116–20
pharmaceutical industry, 67, 148
pharmacogenomics, 25
Plant Patent Act, 185–86
Plant Variety Protection Act, 185–86
Plausibility, 110, 111, 127, 141, 207
    vs. probability, 110, 207
polymerase chain reaction, 27
pragmatics, 87
precautionary principle, 10, 110–12,
    125–26, 131, 152, 167, 192, 196–97
probability, 108–10, 125–26, 196, 207,
    209
product of nature vs. product of human
    ingenuity, 39, 53, 73, 76, 84–89,
    174–75, 187–88, 197
professionalism, 169

property, 32, 64, 65, 76, 101, 104 116,
    157
    intangible, 32, 191–92
    intellectual (see intellectual property)
    tangible, 32
proteins, 18–25, 28, 58, 91, 199, 203
proteome, 19
proteomics, 25
public domain, 38, 40, 77, 198
public good, 77
public vs. private control, 38, 49, 57, 60,
    81, 99, 153, 172–73

rationality, 207–208
Rawls, J., 9, 80, 204
realism, 203
reasonableness, 108, 111–12, 124, 164
    203, 207–208
reductionism. 89–91, 119
religion, 75–77, 97–98, 180
research, 31
    applied vs. basic, 45
    commercial vs. non-commercial, 45,
        149, 199
    exemption, 44, 45, 48, 152, 199
    funding of, 3, 68, 71, 134–35, 145
ribonucleic acid (RNA), 14, 18, 91,
    199
Rifkin, J., 2
risks, 9, 107, 108, 124–25
    reasonable vs. unreasonable, 107, 108

science, 9
    bias in, 144, 162
    goals of , 133
    misconduct in, 144
    objectivity in, 144
    openness in (see openness)
    progress of, 6, 71, 75, 132–40, 152,
        195, 197
Science Citation Index, 136–37
secrecy, 33, 139, 141–43, 199
sickle cell anemia, 16, 20, 26
single nucleotide polymorphisms
    (SNPs), 3, 17, 18, 78, 150
    consortium, 150
slavery, 112–13

slippery slope arguments, 105, 122–24,
127–28, 183
Smith, A., 104, 206
Smith Kline Beecham, 5, 68
social welfare principle, 38
society, 9, 10, 38, 196–97
specialization, 134
stewardship, 79–82

terminator technology, 187–88
theology, 75–77, 83, 97–98, 180
trademarks, 50
Trade Related Aspects of Intellectual
Properties (TRIPS), 60–61, 174, 209
trust, 144, 164

United Nations Educational, Social and
Cultural Organization (UNESCO),
3, 61, 74, 77

University of California, 44, 67
University of Pennsylvania, 161, 167
Utilitarianism, 8, 34, 35, 37, 63, 67–74,
79–80, 88, 186, 204–206

value(s)
intrinsic vs. extrinsic, 95, 205–206
symbolic, 7, 115
Venter, C., 4, 5, 71
vitamin B_{12}, 85
vitalism, 80

Watson, J., 14, 22
Watt, J., 33
Wilmut, I., 179
Wilson, J., 161, 167
World Health Organization (WHO),
61